Bayesian Inference on Complicated Data

Edited by Niansheng Tang

Published in London, United Kingdom

IntechOpen

Supporting open minds since 2005

Bayesian Inference on Complicated Data
http://dx.doi.org/10.5772/intechopen.83214
Edited by Niansheng Tang

Contributors
Ying-Ying Zhang, Hongsheng Dai, Christophe Ley, Fatemeh Ghaderinezhad, Xi Chen, Jianhua Xuan, Catherine C. Liu, Junshan Shen, Michelle Yongmei Wang, Trevor Park, Shahid Naseem

First published in London, United Kingdom, 2020 by IntechOpen
IntechOpen is the global imprint of INTECHOPEN LIMITED, registered in England and Wales, registration number: 11086078, 7th floor, 10 Lower Thames Street, London, EC3R 6AF, United Kingdom
Printed in Croatia

British Library Cataloguing-in-Publication Data
A catalogue record for this book is available from the British Library

Additional hard and PDF copies can be obtained from orders@intechopen.com

Bayesian Inference on Complicated Data
Edited by Niansheng Tang
p. cm.
Print ISBN 978-1-83880-385-8
Online ISBN 978-1-83880-386-5
eBook (PDF) ISBN 978-1-83962-704-0

For EU product safety concerns:
IN TECH d.o.o., Prolaz Marije Krucifikse Kozulić 3, 51000 Rijeka, Croatia, info@intechopen.com or visit our website at intechopen.com.

Meet the editor

Niansheng Tang is Professor of Statistics and dean of the School of Mathematics and Statistics, Yunnan University. He was elected a Yangtze River Scholars Distinguished Professor in 2013, a member of the International Statistical Institute (ISI) in 2016, and a member of the board of International Chinese Statistical Association (ICSA) in 2018. He obtained the National Science Foundation for Distinguished Young Scholars of China in 2012. He serves as a member of the editorial board for *Statistics and Its Interface* and *Journal of Systems Science and Complexity*. He is also an editor for *Communications in Mathematics and Statistics*. His research interests include biostatistics, Bayesian statistics, missing data analysis, statistical diagnosis, variable selection, and high-dimensional data analysis. He has published more than 170 research papers and authored fours books.

Contents

Preface

Over the years, due to great applications in various fields such as social science, biomedicine, genomics, and signal processing, and the improvement of computing ability, Bayesian statistics have made substantial developments. In particular, many novel Bayesian theories and methods, including novel sampling techniques, the selection of the prior, and new Bayesian estimation procedures, have been developed. This book introduces key ideas of Bayesian sampling methods, Bayesian estimation, and the selection of the prior. This book is structured around topics on the impact of the choice of the prior on Bayesian statistics, some advances on Bayesian sampling methods, and Bayesian inference for complicated data including breast cancer data, cloud-based healthcare data, gene network data, and longitudinal data.

Fundamental statistical problems have changed with the move from continuous/discrete data to network and cloud-based data analyses. As a result of network and cloud-based data analyses, traditional Bayesian sampling techniques suffer from unprecedented challenges. To this end, this book introduces some novel approaches to make Bayesian inference on a few topics of interest, rather than give a comprehensive overview.

This book includes three sections and seven chapters. Section I introduces the impact problem of the choice of the prior. It includes Chapter 1, in which Professor Ley Christophe investigates the impact of the choice of the prior on Bayesian statistics including conjugate prior and Jeffrey's prior. Section II focuses on some advances on sampling methods. It contains Chapters 2 and 3, in which Professor Wang Michelle introduces Gibbs sampler, slice sampler, Metropolis-Hastings sampling, Hamiltonian Monte Carlo, and cluster sampling, among others, and Professor Dai Hongsheng reviews exact Monte Carlo simulation techniques. Section III describes Bayesian inference for complicated data. It contains Chapters 4, 5, 6, and 7, in which Professor Liu Catherine introduces Bayesian analysis for random effects models, Professor Chen Xi studies Bayesian integration for gene network data, Professor Nguyen Loc discusses Bayesian inference for cloud-based healthcare data, and Dr. Zhang Ying-Ying considers Bayesian estimators under six loss functions.

I was invited to edit this book after the publication of "Bayesian analysis for hidden Markov factor analysis models," which I co-wrote with Xia Yemao, Zeng Xiaoqian, and Tang Niansheng. I am very grateful to Mr. Mateo Pulko for his kind invitation to edit this book and for providing me the chance to work with my aforementioned coauthors. I would also like to thank Professors Ley Christophe, Wang Michelle, Dai Hongsheng, Liu Catherine, Chen Xi, Nguyen

Loc, and Zhang Ying-Ying for their contributions. I sincerely hope that this book will be of great interest to statisticians, engineers, doctors, and machine learning researchers.

Niansheng Tang
Yunnan University,
China

Section 1

The Choice of the Prior

Chapter 1

On the Impact of the Choice of the Prior in Bayesian Statistics

Fatemeh Ghaderinezhad and Christophe Ley

Abstract

A key question in Bayesian analysis is the effect of the prior on the posterior, and how we can measure this effect. Will the posterior distributions derived with distinct priors become very similar if more and more data are gathered? It has been proved formally that, under certain regularity conditions, the impact of the prior is waning as the sample size increases. From a practical viewpoint it is more important to know what happens at finite sample size n. In this chapter, we shall explain how we tackle this crucial question from an innovative approach. To this end, we shall review some notions from probability theory such as the Wasserstein distance and the popular Stein's method, and explain how we use these a priori unrelated concepts in order to measure the impact of priors. Examples will illustrate our findings, including conjugate priors and the Jeffreys prior.

Keywords: conjugate prior, Jeffreys prior, prior distribution, posterior distribution, Stein's method, Wasserstein distance

1. Introduction

A key question in Bayesian analysis is the choice of the prior in a given situation. Numerous proposals and divergent opinions exist on this matter, but our aim is not to delve into a review or discussion, rather we want to provide the reader with a description of a useful new tool allowing him/her to make a decision. More precisely, we explain how to effectively measure the effect of the choice of a given prior on the resulting posterior. How much do two posteriors, derived from two distinct priors, differ? Providing a quantitative answer to this question is important as it also informs us about the ensuing inferential procedures. It has been proved formally in [1, 2] that, under certain regularity conditions, the impact of the prior is waning as the sample size increases. From a practical viewpoint it is however more interesting to know what happens at finite sample size n, and this is precisely the situation we are considering in this chapter.

Recently, [3, 4] have devised a novel tool to answer this question. They measure the Wasserstein distance between the posterior distributions based on two distinct priors at fixed sample size n. The Wasserstein (more precisely, Wasserstein-1) distance is defined as

$$d_{\mathcal{W}}(P_1, P_2) = \sup_{h \in \mathcal{H}} |\mathrm{E}[h(X_1)] - \mathrm{E}[h(X_2)]|$$

for X_1 and X_2 random variables with respective distribution functions P_1 and P_2, and where $\mathcal{H}$ stands for the class of Lipschitz-1 functions. It is a popular distance between two distributions, related to optimal transport and therefore also known as *earth mover distance* in computer science, see [5] for more information. The resulting distance thus gives us the desired measure of the difference between two posteriors. If one of the two priors is the flat uniform prior (leading to the posterior coinciding with the data likelihood), then this measure quantifies how much the other chosen prior has impacted on the outcome as compared to a data-only posterior. Now, the Wasserstein distance being mostly impossible to calculate exactly, it is necessary to obtain sharp upper and lower bounds, which will partially be achieved by using techniques from the so-called Stein method, a famous tool in probabilistic approximation theory. We opt for the Wasserstein metric instead of, e.g., the Kullback-Leibler divergence because of precisely its nice link with the Stein method, see [3].

The chapter is organized as follows. In Section 2 we provide the notations and terminology used throughout the paper, provide the reader with the minimal necessary background knowledge on the Stein method, and state the main result regarding the measure of the impact of priors. Then in Section 3 we illustrate how this new measure works in practice, by first working out a completely new example, namely priors for the scale parameter of the inverse gamma distribution, and second giving new insights into an example first treated in both [3, 4], namely priors for the success parameter in the binomial distribution.

2. The measure in its most general form

In this section we provide the reader with the general form of the new measure of the impact of the choice of prior distributions. Before doing so, we however first give a very brief overview on Stein's method that is of independent interest.

2.1 Stein's method in a nutshell

Stein's method is a popular tool in applied and theoretical probability, typically used for Gaussian and Poisson approximation problems. The principal goal of the method is to provide quantitative assessments in distributional comparison statements of the form $W \approx Z$ where Z follows a known and well-understood probability distribution (typically normal or Poisson) and W is the object of interest. Charles Stein [6] in 1972 laid the foundation of what is now called "Stein's method" by aiming at normal approximations.

Stein's method consists of two distinct components, namely

Part A: a framework allowing to convert the problem of bounding the error in the approximation of W by Z into a problem of bounding the expectation of a certain functional of W.

Part B: a collection of techniques to bound the expectation appearing in Part A; the details of these techniques are strongly dependent on the properties of W as well as on the form of the functional.

We refer the interested reader to [7, 8] for detailed recent accounts on this powerful method. The reader will understand in the next sections why Stein's method has been of use for quantifying the desired measure, even without formal proofs or mathematical details.

2.2 Notation and formulation of the main goal

We start by fixing our notations. We consider independent and identically distributed (discrete or absolutely continuous) observations $X_1, \ldots, X_n$ from a parametric model with parameter of interest $\theta \in \Theta \subseteq \mathbb{R}$. We denote the likelihood of $X_1, \ldots, X_n$ by $\ell(x; \theta)$ where $x = (x_1, \ldots, x_n)$ are the observed values. Take two different (possibly improper) prior densities $p_1(\theta)$ and $p_2(\theta)$ for our parameter θ; the famous Bayes' theorem then readily yields the respective posterior densities

$$p_i(\theta; x) = \kappa_i(x) p_i(\theta) \ell(x; \theta), \quad i = 1, 2,$$

where $\kappa_1(x)$, $\kappa_2(x)$ are normalizing constants that depend only on the observed values. We denote by (Θ_1, P_1) and (Θ_2, P_2) the couples of random variables and cumulative distribution functions associated with the densities $p_1(\theta; x)$ and $p_2(\theta; x)$.

These notations allow us to formulate the main goal: measure the Wasserstein distance between $p_1(\theta; x)$ and $p_2(\theta; x)$, as this will exactly correspond to the difference between the posteriors resulting from the two priors p_1 and p_2. Sharp upper and lower bounds have been provided for this Wasserstein distance, first in [3] for the special case of one prior being flat uniform, then in all generality in [4]. The determination of the upper bound has been achieved by means of the Stein Method: first a relevant Stein operator has been found (Part A), and then a new technique designed in [3] has been put to use for Part B. The reader is referred to these two papers for details about the calculations; since this chapter is part of a book on Bayesian inference, we prefer to keep out those rather probabilistic manipulations.

2.3 The general result

The key element in the mathematical developments underlying the present problem is that the densities $p_1(\theta; x)$ and $p_2(\theta; x)$ are *nested*, meaning that one support is included in the other. Without loss of generality we here suppose that $I_2 \subseteq I_1$, allowing us to express $p_2(\theta; x)$ as $\frac{\kappa_2(x)}{\kappa_1(x)} \rho(\theta) p_1(\theta; x)$ with

$$\rho(\theta) = \frac{p_2(\theta)}{p_1(\theta)}.$$

The following general result has been obtained in [4], where we refer the reader to for a proof.

Theorem 1.1 Consider $\mathcal{H}$ the set of Lipschitz-1 functions on $\mathbb{R}$ and define

$$\tau_i(\theta; x) = \frac{1}{p_i(\theta; x)} \int_{a_i}^{\theta} (\mu_i - y) p_i(y; x) dy, \quad i = 1, 2, \tag{1}$$

where a_i is the lower bound of the support $I_i = (a_i, b_i)$ of p_i. Suppose that both posterior distributions have finite means μ_1 and μ_2, respectively. Assume that $\theta \mapsto \rho(\theta)$ is differentiable on I_2 and satisfies (i) $\mathrm{E}[|\Theta_1 - \mu_1| \rho(\Theta_1)] < \infty$, (ii) $\left(\rho(\theta) \int_{a_1}^{\theta} (h(y) - \mathrm{E}[h(\Theta_1)]) p_1(y; x) dy\right)$ is integrable for all $h \in \mathcal{H}$ and (iii) $\lim_{\theta \to a_2, b_2} \rho(\theta) \int_{a_1}^{\theta} (h(y) - \mathrm{E}[h(\Theta_1)]) p_1(y; x) dy = 0$ for all $h \in \mathcal{H}$. Then

$$|\mu_1 - \mu_2| = \frac{|\mathrm{E}[\tau_1(\Theta_1; x) \rho'(\Theta_1)]|}{\mathrm{E}[\rho(\Theta_1)]} \leq d_{\mathcal{W}}(P_1, P_2) \leq \frac{\mathrm{E}[\tau_1(\Theta_1; x) |\rho'(\Theta_1)|]}{\mathrm{E}[\rho(\Theta_1)]}$$

and, if the variance of Θ_1 exists,

$$|\mu_1 - \mu_2| \leq d_{\mathcal{W}}(P_1, P_2) \leq \|\rho'\|_\infty \frac{Var[\Theta_1]}{\mathrm{E}[\rho(\Theta_1)]}$$

where $\|\cdot\|_\infty$ stands for the infinity norm.

This result quantifies in all generality the measure of the difference between two priors p_1 and p_2, and comprises of course the special case where one prior is flat uniform. Quite nicely, if ρ is a monotone increasing or decreasing function, the bounds do coincide, leading to

$$d_{\mathcal{W}}(P_1; P_2) = \frac{\mathrm{E}[\tau_1(\Theta_1; x)|\rho'(\Theta_1)|]}{\mathrm{E}[\rho(\Theta_1)]}, \tag{2}$$

hence an exact result. The reader notices the sharpness of these bounds given that they contain the same quantities in both the upper and lower bounds; this fact is further underpinned by the equality Eq. (2). Finally we wish to stress that the functions $\tau_i(\theta; x), i = 1, 2,$ from Eq. (1) are called Stein kernel in the Stein method literature and that these functions are always positive and vanish at the boundaries of the support.

3. Applications and illustrations

Numerous examples have been treated in [3, 4], such as priors for the location parameter of a normal distribution, the scale parameter of a normal distribution, the success parameter of a binomial or the event-enumerating parameter of the Poisson distribution, to cite but these. In this section we will, on the one hand, investigate a new example, namely the scale parameter of an inverse gamma distribution, and, on the other hand, revisit the binomial case. Besides providing the bounds, we will also for the first time plot numerical values for the bounds and hence shed new intuitive light on this measure of the impact of the choice of the prior.

3.1 Priors for the scale parameter of the inverse gamma (IG) distribution

The inverse gamma (IG) distribution has the probability density function

$$x \to \frac{\beta^\alpha}{\Gamma(\alpha)} x^{-\alpha-1} \exp\left\{-\frac{\beta}{x}\right\}, \; x>0,$$

where α and β are the positive shape and scale parameters, respectively. This distribution corresponds to the reciprocal of a gamma distribution (if $X \sim Gamma(\alpha, \beta)$ then $\frac{1}{X} \sim IG(\alpha, \beta)$) and is frequently encountered in domains such as machine learning, survival analysis and reliability theory. Within Bayesian Inference, it is a popular choice as prior for the scale parameter of a normal distribution. In the present setting, we consider $\theta = \beta$ as the parameter of interest and α is fixed. The observations sampled from this distribution are written $x_1, \ldots, x_n$.

The first prior is the popular noninformative Jeffreys prior. It is invariant under reparameterization and is proportional to the square root of the Fisher information quantity associated with the parameter of interest. In the present setting simple calculations show that it is proportional to $\frac{1}{\beta}$. The resulting posterior P_1 then has a density of the form

$$p_1(\beta|x) \propto \frac{1}{\beta}\beta^{n\alpha} \exp\left\{-\beta\sum_{i=1}^{n}\frac{1}{x_i}\right\} = \beta^{n\alpha-1}\exp\left\{-\beta\sum_{i=1}^{n}\frac{1}{x_i}\right\}$$

which is none other than a gamma distribution with parameters $\left(n\alpha, \sum_{i=1}^{n}\frac{1}{x_i}\right)$.

Now, the gamma distribution happens to be the conjugate prior for the scale parameter of an IG distribution. We consider thus as second prior a general gamma distribution with density $\beta \mapsto \frac{\kappa^\eta}{\Gamma(\eta)}\beta^{\eta-1}\exp\{-\kappa\beta\}$, where the shape and scale parameters η and κ are strictly positive. The ensuing posterior distribution P_2 has then the density

$$p_2(\beta|x) \propto \beta^{\eta-1}\exp\{-\kappa\beta\} \times \beta^{n\alpha}\exp\left\{-\beta\sum_{i=1}^{n}\frac{1}{x_i}\right\} = \beta^{n\alpha+\eta-1}\exp\left\{-\beta\left(\sum_{i=1}^{n}\frac{1}{x_i}+\kappa\right)\right\}$$

which is a gamma distribution with updated parameters $\left(n\alpha+\eta, \sum_{i=1}^{n}\frac{1}{x_i}+\kappa\right)$.

Considering Jeffreys prior as p_1 and the gamma prior as p_2 leads to the ratio

$$\rho(\beta) = \frac{p_2(\beta)}{p_1(\beta)} \propto \frac{\frac{\kappa^\eta}{\Gamma(\eta)}\beta^{\eta-1}\exp\{-\kappa\beta\}}{\frac{1}{\beta}} = \frac{\kappa^\eta}{\Gamma(\eta)}\beta^\eta\exp\{-\kappa\beta\}.$$

One can easily check that all conditions of Theorem 1.1 are fulfilled, hence we can calculate the bounds. The lower bound is directly obtained as follows:

$$d_{\mathcal{W}}(P_1,P_2) \geq |\mu_1-\mu_2| = \left|\frac{n\alpha}{\sum_{i=1}^{n}\frac{1}{x_i}} - \frac{n\alpha+\eta}{\sum_{i=1}^{n}\frac{1}{x_i}+\kappa}\right| \tag{3}$$

$$= \left|\frac{n\alpha\sum_{i=1}^{n}\frac{1}{x_i}+n\alpha\kappa-n\alpha\sum_{i=1}^{n}\frac{1}{x_i}-\eta\sum_{i=1}^{n}\frac{1}{x_i}}{\sum_{i=1}^{n}\frac{1}{x_i}\left(\sum_{i=1}^{n}\frac{1}{x_i}+\kappa\right)}\right| \tag{4}$$

$$= \left|\frac{n\alpha\kappa-\eta\sum_{i=1}^{n}\frac{1}{x_i}}{\sum_{i=1}^{n}\frac{1}{x_i}\left(\sum_{i=1}^{n}\frac{1}{x_i}+\kappa\right)}\right|. \tag{5}$$

In order to acquire the upper bound we need to calculate

$$\rho'(\beta) = \frac{\kappa^\eta}{\Gamma(\eta)}\beta^{\eta-1}\exp(-\kappa\beta)[\eta-\kappa\beta]$$

and, writing Θ_1 the random variable associated with $Gamma\left(n\alpha, \sum_{i=1}^{n}\frac{1}{x_i}\right)$ and $f_{Gamma\left(n\alpha, \sum_{i=1}^{n}\frac{1}{x_i}\right)}(\beta)$ the related density, we get

$$\mathbb{E}[\rho(\Theta_1)] = \int_0^\infty \frac{\kappa^\eta}{\Gamma(\eta)}\beta^\eta\exp\{-\kappa\beta\} \times f_{Gamma\left(n\alpha, \sum_{i=1}^{n}\frac{1}{x_i}\right)}(\beta)d\beta \tag{6}$$

$$= \frac{\kappa^\eta}{\Gamma(\eta)}\frac{\left(\sum_{i=1}^{n}\frac{1}{x_i}\right)^{n\alpha}}{\Gamma(n\alpha)}\int_0^\infty \beta^\eta\exp\{-\kappa\beta\}\beta^{n\alpha-1}\exp\left\{-\beta\sum_{i=1}^{n}\frac{1}{x_i}\right\}d\beta \tag{7}$$

$$= \frac{\kappa^{\eta}}{\Gamma(\eta)} \frac{\left(\sum_{i=1}^{n} \frac{1}{x_i}\right)^{n\alpha}}{\Gamma(n\alpha)} \int_0^{\infty} \beta^{n\alpha+\eta-1} \exp\left\{-\beta\left(\sum_{i=1}^{n} \frac{1}{x_i} + \kappa\right)\right\} d\beta \tag{8}$$

$$= \frac{\kappa^{\eta}}{\Gamma(\eta)} \frac{\left(\sum_{i=1}^{n} \frac{1}{x_i}\right)^{n\alpha}}{\Gamma(n\alpha)} \frac{\Gamma(n\alpha+\eta)}{\left(\sum_{i=1}^{n} \frac{1}{x_i} + \kappa\right)^{n\alpha+\eta}} \tag{9}$$

$$= \frac{\kappa^{\eta}}{Beta(n\alpha, \eta)} \frac{\left(\sum_{i=1}^{n} \frac{1}{x_i}\right)^{n\alpha}}{\left(\sum_{i=1}^{n} \frac{1}{x_i} + \kappa\right)^{n\alpha+\eta}}. \tag{10}$$

From the Stein literature we know that the Stein kernel for the gamma distribution with parameters $\left(n\alpha, \sum_{i=1}^{n} \frac{1}{x_i}\right)$ corresponds to $\tau(\beta;x) = \frac{\beta}{\sum_{i=1}^{n} \frac{1}{x_i}}$. Employing the triangular inequality we have thus

$$\mathbb{E}[\tau(\Theta_1; x)|\rho'(\Theta_1)|] = \mathbb{E}\left[\frac{\Theta_1}{\sum_{i=1}^{n} \frac{1}{x_i}} \frac{\kappa^{\eta}}{\Gamma(\eta)} \Theta_1^{\eta-1} \exp\{-\kappa\Theta_1\}|\eta - \kappa\Theta_1|\right] \tag{11}$$

$$\leq \frac{\kappa^{\eta}}{\left(\sum_{i=1}^{n} \frac{1}{x_i}\right)\Gamma(\eta)} \mathbb{E}\left[\Theta_1^{\eta} \exp\{-\kappa\Theta_1\}(\eta + \kappa\Theta_1)\right]. \tag{12}$$

Now we need to calculate the expectation

$$\mathbb{E}\left[\Theta_1^{\eta} \exp\{-\kappa\Theta_1\}(\eta + \kappa\Theta_1)\right] \tag{13}$$

$$= \int_0^{\infty} \beta^{\eta} \exp\{-\kappa\beta\}(\eta + \kappa\beta) \times f_{Gamma\left(n\alpha, \sum_{i=1}^{n} \frac{1}{x_i}\right)}(\beta) d\beta \tag{14}$$

$$= \frac{\left(\sum_{i=1}^{n} \frac{1}{x_i}\right)^{n\alpha}}{\Gamma(n\alpha)} \int_0^{\infty} \eta\beta^{n\alpha+\eta-1} \exp\left\{-\beta\left(\sum_{i=1}^{n} \frac{1}{x_i} + \kappa\right)\right\} d\beta \tag{15}$$

$$+ \frac{\left(\sum_{i=1}^{n} \frac{1}{x_i}\right)^{n\alpha}}{\Gamma(n\alpha)} \int_0^{\infty} \kappa\beta^{n\alpha+\eta} \exp\left\{-\beta\left(\sum_{i=1}^{n} \frac{1}{x_i} + \kappa\right)\right\} d\beta \tag{16}$$

$$= \frac{\left(\sum_{i=1}^{n} \frac{1}{x_i}\right)^{n\alpha}}{\Gamma(n\alpha)} \left(\eta \frac{\Gamma(n\alpha+\eta)}{\left(\sum_{i=1}^{n} \frac{1}{x_i} + \kappa\right)^{n\alpha+\eta}} + \kappa \frac{\Gamma(n\alpha+\eta+1)}{\left(\sum_{i=1}^{n} \frac{1}{x_i} + \kappa\right)^{n\alpha+\eta+1}}\right). \tag{17}$$

The final expression for the upper bound then corresponds to

$$d_{\mathcal{W}}(P_1, P_2) \leq \frac{\frac{\kappa^{\eta}}{\left(\sum_{i=1}^{n} \frac{1}{x_i}\right)\Gamma(\eta)} \times \frac{\left(\sum_{i=1}^{n} \frac{1}{x_i}\right)^{n\alpha}}{\Gamma(n\alpha)} \left[\eta \frac{\Gamma(n\alpha+\eta)}{\left(\sum_{i=1}^{n} \frac{1}{x_i}+\kappa\right)^{n\alpha+\eta}} + \kappa \frac{\Gamma(n\alpha+\eta+1)}{\left(\sum_{i=1}^{n} \frac{1}{x_i}+\kappa\right)^{n\alpha+\eta+1}}\right]}{\frac{\kappa^{\eta}}{Beta(n\alpha, \eta)} \times \frac{\left(\sum_{i=1}^{n} \frac{1}{x_i}\right)^{n\alpha}}{\left(\sum_{i=1}^{n} \frac{1}{x_i}+\kappa\right)^{n\alpha+\eta}}} \tag{18}$$

$$= \frac{Beta(n\alpha, \eta)\left(\sum_{i=1}^{n} \frac{1}{x_i} + \kappa\right)^{n\alpha+\eta}}{\frac{\Gamma(n\alpha)\Gamma(\eta)}{\Gamma(n\alpha+\eta)}\left(\sum_{i=1}^{n} \frac{1}{x_i}\right)} \times \frac{1}{\left(\sum_{i=1}^{n} \frac{1}{x_i} + \kappa\right)^{n\alpha+\eta}} \left(\eta + \kappa \frac{n\alpha+\eta}{\sum_{i=1}^{n} \frac{1}{x_i} + \kappa}\right) \tag{19}$$

$$= \frac{1}{\sum_{i=1}^{n} \frac{1}{x_i}} \left(\eta + \kappa \frac{n\alpha + \eta}{\sum_{i=1}^{n} \frac{1}{x_i} + \kappa} \right). \tag{20}$$

The Wasserstein distance between the posteriors based on the Jeffreys prior and conjugate gamma prior for the scale parameter β of the IG distribution is thus bounded as

$$\left| \frac{n\alpha\kappa - \eta \sum_{i=1}^{n} \frac{1}{x_i}}{\left(\sum_{i=1}^{n} \frac{1}{x_i}\right)\left(\sum_{i=1}^{n} \frac{1}{x_i} + \kappa\right)} \right| \le d_{\mathcal{W}(P_1, P_2)} \le \frac{1}{\sum_{i=1}^{n} \frac{1}{x_i}} \left(\eta + \kappa \frac{n\alpha + \eta}{\kappa + \sum_{i=1}^{n} \frac{1}{x_i}} \right).$$

It can be seen that both the lower and upper bound are of the order of $O(n^{-1})$. In addition, it is noticeable that for the larger observations, the rate of convergence is getting slower.

In order to show the performance of the methodology which leads to have the lower and upper bounds, we have conducted a simulation study including two parts. First we simulate $N = 100$ samples for each sample size $n = 10, 11, \cdots, 100$ from the inverse gamma distribution with parameters $(\alpha, \beta) = (0.5, 1)$ in each iteration. For each of these samples we calculate the lower and upper bounds of the Wasserstein distance and calculate the average over all N replications, together with the difference between the bounds. Finally we plot these values for each sample size in **Figure 1**. We repeat the same process for $N = 1000$ samples with the same sizes. The hyperparameters from the prior gamma distribution are $(\kappa, \eta) = (0.2, 2)$. We clearly observe how fast these values decrease with the sample size. Of course, augmenting the number of replications does not increase the speed of convergence, however the curves become noticeably smoother.

This methodology not only can help the practitioners to make a decision between existing priors in theory, but also helps them to know from what sample size on the effect of choosing one prior becomes less important, especially in situations when the cost and time matter. This can be particularly useful when the

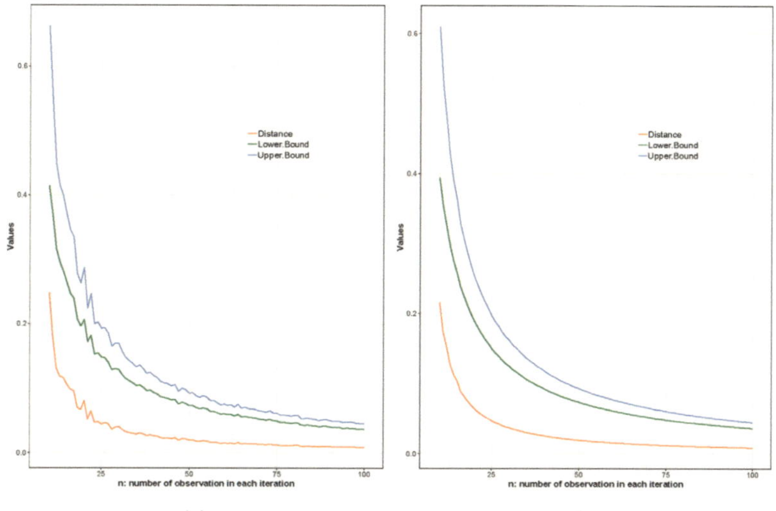

Figure 1.
(a) Shows the bounds and the distances between the bounds for N = 100 iterations for each sample size 10–100 by steps of 1, and (b) illustrates the same situation for N = 1000. The hyperparameters are $\kappa = 0.2$ and $\eta = 2$, while the fixed parameter α equals 0.5.

hesitation is between a simple, closed-form prior and a more complicated one. It is advisable to use the simpler one when there is no considerable difference between the effect of the two priors.

3.2 The impact of priors for the success parameter of the binomial model

The probability mass function of a binomial distribution is given by

$$x \mapsto \binom{n}{x} \theta^x (1-\theta)^{n-x}$$

where $x \in \{0, 1, \cdots, n\}$ is the number of observed successes, the natural number n indicates the number of binary trials and $\theta \in (0, 1)$ stands for the success parameter. In this setting we suppose n is fixed and the underlying parameter of interest is θ.

A comprehensive comparison of various priors for the binomial distribution including a beta prior, the Haldane prior and Jeffreys prior, has been done in [9], based on the methodology described above. Therefore, since there is a complete reference for the reader in this case, we use the binomial distribution as a second example to show numerical results.

The theoretical lower and upper bounds between a $Beta(\alpha, \beta)$ prior and the flat uniform prior are given by

$$\left| \frac{x+1}{n+2} \left(\frac{\alpha+\beta-2}{n+\alpha+\beta} \right) - \frac{\alpha-1}{n+\alpha+\beta} \right| \leq d_{\mathcal{W}}(P_1, P_2) \leq \frac{1}{n+2} \left\{ |\alpha-1| + \frac{x+\alpha}{n+\alpha+\beta} (|\beta-1| - |\alpha-1|) \right\},$$

where x is the observed number of successes. We see that both lower and upper bounds are of the order of $O(n^{-1})$. This rate of convergence remains even in the extreme cases $x = 0$ and $x = n$. We invite the reader to see [3, 9] for more details.

In order to illustrate the behavior of the lower and upper bounds and the distances between them, we have conducted a two-part simulation study for the binomial distribution. First, we consider 100 sample sizes (number of trials in the binomial distribution) varying from 10 to 1000 by steps of 10, and generate binomial data exactly once for every sample size (with $\theta = 0.2$). The results of the bounds, obtained for hyperparameters $(\alpha, \beta) = (2, 4)$ from the beta prior, are reported in **Figure 2a** and we can see that, even with only one iteration, when the number of trials (the sample size) increases the lower and upper bound become closer, which is a numerical quantification of the fact that the influence of the choice of the prior wanes asymptotically. This becomes also visible from the distance between the two bounds. Sampling only once for each sample size leads to slightly unpleasant variations in the lower bounds (non-monotone behavior), which however nearly disappear in the second considered scenario. Indeed, in **Figure 2b** we increased the number of iterations to 50 for the same different sample sizes and took averages. A better smoothness is the consequence. This simulation study not only provides the reader with numerical values for the bounds, to which he/she can compare his/her bounds obtained for real data, but also gives a nice visualization of the impact of the choice of the prior at fixed sample size. The main conclusion is that the impact drops fast at small sample sizes, and the bounds start to become very close for medium-to-large sample sizes.

Finally, we investigate the impact of the hyperparameters on the upper and lower bounds. To this end, we varied both α and β in **Table 1**. The situation with α fixed to two and relatively small β corresponds well with $p = 0.2$, which explains why the upper and lower bounds, and hence the Wasserstein distance and thus the impact of the prior, are the smallest. Increasing β more augments the distance.

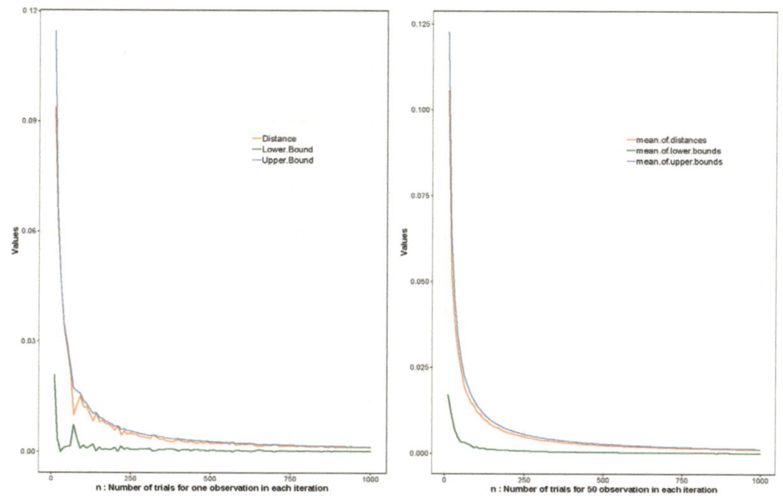

(a) (b)

Figure 2.
(a) Shows the lower and upper bounds and the distances for the number of trials $\{n = 10,\ldots,1000\}$ for one iteration. (b) Shows the same situation, however this time based on averages obtained for 50 iterations. In both situations the hyperparameters from the beta prior are $\alpha = 2$ and $\beta = 4$.

Hyperparameters (α, β)	Average of the lower bounds	Average of the upper bounds
(0.2, 0.4)	0.002561383	0.003726728
(0.2, 0.8)	0.00296002	0.003344393
(2, 2)	0.002699325	0.00490119
(2, 5)	0.0008115384	0.007984289
(2, 10)	0.004506271	0.01241273
(2, 15)	0.008208887	0.01626326
(2, 30)	0.01750177	0.02581062
(2, 50)	0.02739205	0.0359027
(2, 100)	0.04592235	0.05470826
(2, 200)	0.07071766	0.07976386
(2, 500)	0.1103048	0.1196464
(2, 1000)	0.1399961	0.1495087
(10, 2)	0.02813367	0.03132908
(35, 2)	0.08571115	0.09033568
(50, 2)	0.1127136	0.1178113
(100, 2)	0.1830272	0.189071
(200, 2)	0.2783722	0.2853418
(400, 2)	0.3933338	0.401145
(700, 2)	0.4901209	0.4985089
(1000, 2)	0.5482869	0.5569829

Table 1.
The summary of upper and lower bounds for different hyperparameters, with $p = 0.2$ and for $N = 50$ iterations.

(α, β)	$(2, 5)$
Lower Bound (L)	0.0008115384
Upper Bound (U)	0.007984289
Distance (U-L)	0.007172751

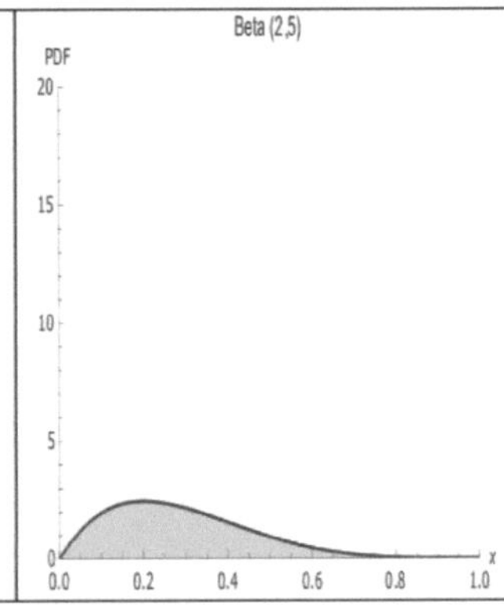

(a)

(α, β)	$(2, 10)$
Lower Bound (L)	0.004506271
Upper Bound (U)	0.01241273
Distance (U-L)	0.007906458

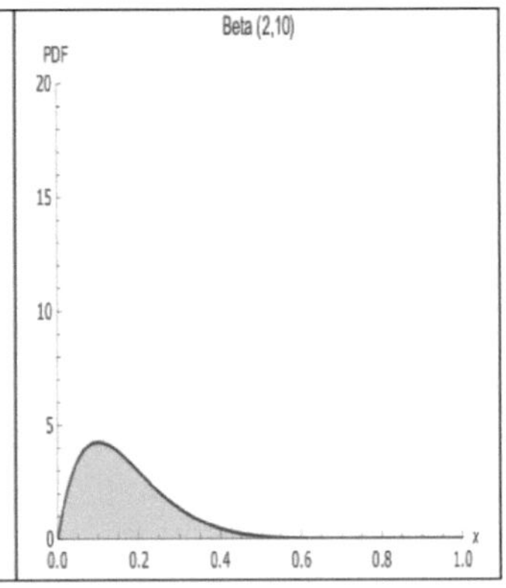

(b)

(α, β)	$(50, 2)$
Lower Bound (L)	0.1127136
Upper Bound (U)	0.1178113
Distance (U-L)	0.00509779

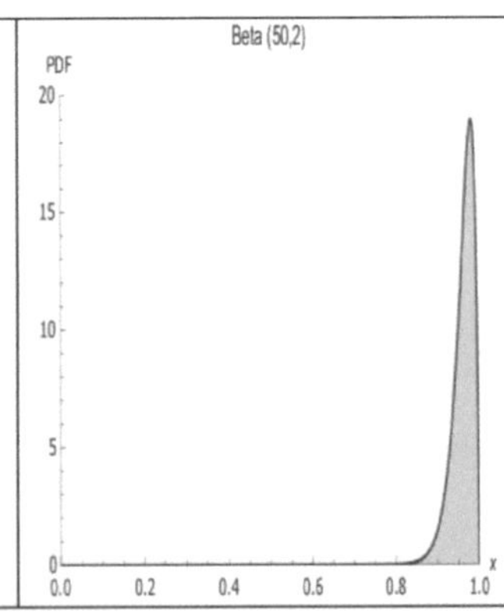

(c)

Figure 3.
Plots of the beta prior densities together with the average lower and upper bounds (and their difference) on the Wasserstein distance between the data-based posterior and the posterior resulting from each beta prior.

On the contrary, fixing $\beta = 2$ yields priors rather centered around large values of p and hence bigger distances. Moreover, the more α is increased, the more the distance augments, as the prior is further away from the data and hence impacts more on the posterior at a fixed sample size. For the sake of illustration, we present three choices of hyperparameters together with the bounds and the related prior density in **Figure 3**. This will help understanding our conclusions.

4. Conclusions

In this chapter we have presented a recently developed measure for the impact of the choice of the prior distribution in Bayesian statistics. We have presented the general theoretical result, explained how to use it in a particular example and provided some graphics to illustrate it numerically. The practical importance of this study is when practitioners hesitate between two proposed priors in a given situation. For instance, Kavetski et al. [10] considered a storm depth multiplier model to represent rainfall uncertainty where the errors appear under multiplicative form and are assumed to be normal. They fix the mean, but state that "less is understood about the degree of rainfall uncertainty," i.e., the multiplier variance, and therefore studied various priors for the variance. Knowledge of the tools presented in this chapter would have simplified the decision process.

In case of missing data, the present methodology can still be used. Either the data get imputed, in which case nothing changes, or the missing data simply are left out from the calculation of upper and lower bounds, whose expression does of course not alter.

Further developments on this new measure might lead to a more concrete quantification of words such as "informative, weakly informative, noninformative" priors, and we hope to have stimulated interest in this promising new line of research within Bayesian Inference.

Acknowledgements

This research is supported by a BOF Starting Grant of Ghent University.

Author details

Fatemeh Ghaderinezhad and Christophe Ley*
Department of Applied Mathematics, Computer Science and Statistics, Ghent University, Ghent, Belgium

*Address all correspondence to: christophe.ley@ugent.be

References

[1] Diaconis F, Freedman D. On the consistency of Bayes estimates (with discussion and rejoinder by the authors). The Annals of Statistics. 1986; **14**:1-67

[2] Diaconis F, Freedman D. On inconsistent Bayes estimates of location. The Annals of Statistics. 1986;**14**:68-87

[3] Ley C, Reinert G, Swan Y. Distances between nested densities and a measure of the impact of the prior in Bayesian statistics. Annals of Applied Probability. 2017;**27**:216-241

[4] Ghaderinezhad F, Ley C. Quantification of the impact of priors in Bayesian statistics via Stein's method. Statistics & Probability Letters. 2019; **146**:206-212

[5] Rüschendorf L. Wasserstein metric. In: Michiel H, editor. Encyclopedia of Mathematics. Netherlands: Springer Science+Business Media B.V./Kluwer Academic Publishers; 2001

[6] Stein C. A bound for the error in the normal approximation to the distribution of a sum of dependent random variables. In: Proceedings of the Sixth Berkeley Symposium on Mathematical Statistics and Probability, Univ. California, Berkeley, CA, 1970/ 1971. 1972. pp. 583-602

[7] Ross N. Fundamentals of Stein's method. Probability Surveys. 2011;**8**: 210-293

[8] Ley C, Reinert G, Swan Y. Stein's method for comparison of univariate distributions. Probability Surveys. 2017; **14**:1-52

[9] Ghaderinezhad F. New insights into the impact of the choice of the prior for the success parameter of binomial distributions. Journal of Mathematics, Statistics and Operations Research, forthcoming

[10] Kavetski D, Kuczera G, Franks SW. Bayesian analysis of input uncertainty in hydrological modeling: 1. Theory. Water Resources Research. 2006;**42**:W03407

Section 2

Some Advances on Sampling Methods

Chapter 2

A Brief Tour of Bayesian Sampling Methods

Michelle Y. Wang and Trevor Park

Abstract

Unlike in the past, the modern Bayesian analyst has many options for approximating intractable posterior distributions. This chapter briefly summarizes the class of posterior sampling methods known as Markov chain Monte Carlo, a type of dependent sampling strategy. Varieties of algorithms exist for constructing chains, and we review some of them here. Such methods are quite flexible and are now used routinely, even for relatively complicated statistical models. In addition, extensions of the algorithms have been developed for various goals. General-purpose software is currently also available to automate the construction of samplers, freeing the analyst to focus on model formulation and inference.

Keywords: Markov chain Monte Carlo, Gibbs sampler, slice sampler, Metropolis-Hastings, Hamiltonian Monte Carlo, cluster sampling, JAGS, Stan

1. Introduction

Modern Bayesian data analysis is enabled by specialized computational tools. Except in relatively simple models, explicit solutions for quantities relevant to Bayesian inference are not available. This limitation has sparked the development of many different approximation methods.

Some approximation methods, such as Laplace approximation [1] and variational Bayes [2], are based on replacing the Bayesian posterior density with a computationally convenient approximation. Such methods may have the advantage of relatively quick computation and scalability, but they leave open the question of how much the resulting approximate Bayesian inference can be trusted to reflect the actual Bayesian inference. There is an inherent bias in the approximation that generally cannot be reduced by applying more intensive computation.

When accuracy is important, simulation-based (stochastic) methods offer an attractive alternative. The goal of these methods is to produce a simulation sample (though not necessarily an independent one) from the (joint) posterior distribution. A simulation sample can be used to approximate almost any quantity relevant to Bayesian inference, including posterior expectations, variances, quantiles, and marginal densities. Since the approximations become more exact as more samples are used, accuracy tends to be limited only by the computational resources available.

Random variates from a general probability distribution that has a known density may be simulated using many classical methods, such as accept/reject and importance sampling. However, such methods tend to be efficient only in special cases and often require analytical insight to improve efficiency. The past three decades have seen interest dramatically increase in the category of Markov chain Monte Carlo (MCMC) methods. Unlike most classical methods, MCMC can often be efficiently automated, even for moderately complicated models. A variety of MCMC methods are available, giving the analyst flexibility in implementation. Moreover, general software is now available that automates most computational details, allowing the analyst to focus on model formulation and inference.

The purpose of this chapter is to offer an introduction to Bayesian simulation methods, with emphasis on MCMC. The motivation and popularity of posterior sampling are illustrated in Section 2. Section 3 describes MCMC and the associated issues including convergence monitoring, mixing, and thinning. Varieties of specific sampling methods are provided in Sections 4 and 5, with the general-purpose software implementing them described in Section 6.

2. Posterior sampling

Bayesian inference requires access to the posterior distribution. Let y denote all of the data to be modeled, and suppose its sampling distribution is in a parametric family with density $\pi(y|\theta)$, where θ represents the parameter (usually a vector), including any hyperparameters. If the prior on θ has density $\pi(\theta)$ then, according to Bayes' rule, the posterior distribution has density

$$\pi(\theta|y) = \frac{\pi(y|\theta)\pi(\theta)}{\pi(y)} \propto_{\theta} \pi(y|\theta)\,\pi(\theta) \tag{1}$$

where the proportionality is in θ (not y). (An improper prior can be used, provided the posterior is proper.) The normalizing constant $\pi(y)$ is notoriously difficult to compute, so methods that avoid using it are preferred.

Since $\pi(y|\theta)$ and $\pi(\theta)$ are typically specified by the analyst, the (unnormalized) posterior density is readily available and typically easy to evaluate. Nonetheless, most quantities used in Bayesian inference (posterior expected values, quantiles, marginal densities, etc.) are defined by integrals involving the posterior density, which are usually intractable and are difficult to deterministically approximate when θ has more than a few components.

This explains the popularity of posterior sampling. Given a sample from the posterior of sufficient effective size, posterior expected values can be approximated by sample means, posterior quantiles by sample quantiles, posterior marginal densities by sample-based density estimates, and so forth. Most posterior inference is readily accomplished if an efficient method of sampling from the posterior is available.

Independent sampling from the posterior is seemingly ideal, since relatively few samples are required to obtain a good approximation in most cases, and the approximation error is relatively easy to characterize. Unfortunately, methods for independent sampling have proven difficult to implement in a general way that efficiently scales with the dimension of θ. For example, rejection sampling (accept/reject) is efficient only if the posterior is tightly bounded by a known function proportional to a density that is easy to sample. Finding such a function is generally difficult, and even adaptive variants struggle in high-dimensional situations.

Currently, the most efficient generally adaptable methods use *dependent* sampling. Dependent sampling usually incurs a computational cost of acquiring a larger number of samples to attain a given accuracy, but the flexibility of these methods and their scalability to higher dimensions offset this disadvantage.

3. Sampling with Markov chains

MCMC is a type of dependent sampling in which the samples are obtained from successive states of a discrete-time Markov chain [3]. The Markov chain is designed to be easy to simulate and intended to (eventually) produce samples that have a distribution arbitrarily close to the posterior distribution.

Specifically, the Markov chain is designed to have a particular *stationary distribution*: a distribution on the state space of the chain that is preserved by the transition kernel. If the chain is started in the stationary distribution, all successive states will have the stationary distribution. In the most basic case, the state space will be the range of θ, and the stationary distribution will be the posterior distribution. A collection of successive states can then be regarded as a (dependent) sample from the posterior.

Since starting the chain in the stationary distribution is difficult, MCMC relies on the stationary distribution also being the (unique) *limiting distribution*: the distribution to which the states converge (in law) as the time index increases. Conditions under which the chain converges are technical (e.g., [4]) and can be difficult to verify analytically in complicated models. Thus, though convergence properties may benefit from following some general guidelines in specifying the MCMC method, convergence is usually checked empirically.

General convergence monitoring tools and techniques are available to determine by what time point convergence has been practically achieved, so that accurate samples can be collected thereafter. See [5] for an overview. Some tools rely on simulating the chain several times, independently, from different starting points.

Running the chain(s) until declaring convergence is called *burn-in*, or sometimes *warm up*. All values generated during burn-in are discarded, except for the final state, which becomes the starting point for sampling.

The degree of dependence within a Markov chain determines the number of samples needed for a given level of approximation. Most MCMC methods produce chains with positive dependence, requiring a larger number of samples to be taken than if independent sampling were used. Chains that are highly dependent exhibit slow *mixing*: the decay rate of dependence between the states of the chain at two time points as the time lag increases. In extreme cases, slow mixing makes MCMC computationally prohibitive, since an enormous number of samples may be needed to achieve a reasonable approximation. Methods with fast mixing are typically preferred.

When sampling is highly dependent, using only a regularly spaced subsample of the generated values may be almost as accurate as using all of the values. Retaining only the regularly spaced subsample is called *thinning*. Although it does not reduce the amount of computation required, it can dramatically reduce the time and space required for storage of the values.

Characterizing Monte Carlo error in approximations from an MCMC sample is more difficult than from an independent sample. However, effective methods are available for most cases. See [6].

4. Constructing Markov chains for sampling

This section briefly summarizes the most practical and frequently used methods for forming a Markov Chain appropriate for sampling from a posterior distribution. All of them need only a function proportional to the posterior density of θ, as in Eq. (1). For brevity, we denote it as

$$f(\theta) \propto_\theta \pi(y|\theta)\,\pi(\theta) \tag{2}$$

where the proportionality is in θ only, and the dependence on y has been suppressed in the notation.

4.1 Gibbs sampling

Consider a partition of θ into K pieces (which may themselves be vectors):

$$\theta = (\theta_1, \dots, \theta_K) \tag{3}$$

The *full conditional* (or *conditional posterior*) distribution of θ_k is its posterior distribution conditional on all the other pieces θ_{-k}, i.e., the distribution with density

$$\pi(\theta_k|\theta_{-k}, y) \tag{4}$$

Gibbs sampling, in its purest form, is sequential sampling from the full conditional distributions of θ_k, $k = 1, \dots, K$, each time conditioning upon the most recently sampled value for each component of θ_{-k}. Each complete cycle of this process produces a single sampled value of θ, and these successive values form a Markov chain whose stationary distribution (if unique) is the posterior distribution (since each step in the cycle preserves the posterior distribution of θ).

Essentially, Gibbs sampling reduces the problem of sampling θ to the problem of conditionally sampling each of its pieces. It relies on each full conditional being easy to sample. Because the pieces are of lower dimension (perhaps even one-dimensional), they may be easier to sample by conventional methods. Moreover, it is often possible to choose a prior distribution such that many of the full conditionals are easy to sample. For example, when conditional priors are chosen from easily sampled families that are *partially conjugate* to the sampling model (see, e.g., [7]), the Gibbs sampler is easy to construct. Even if a full conditional cannot be directly sampled, its density is proportional to $f(\theta)$, since

$$\pi(\theta_k|\theta_{-k}, y) = \frac{\pi(\theta|y)}{\pi(\theta_{-k}|y)} \propto_{\theta_k} f(\theta) \tag{5}$$

where the proportionality is in θ_k only (for fixed θ_{-k}). The density of the full conditional is therefore known (up to a constant scaling), so techniques described in the following subsections may be used.

Performance of Gibbs sampling can sometimes be improved by modifying the algorithm. For example, the order in which the pieces are sampled can affect the mixing rate (e.g., [8]). Also, replacing some of the full conditional distributions with (partial) posterior marginals results in a *partially collapsed* Gibbs sampler, which may have better sampling properties [9], though must be implemented carefully to preserve the stationary distribution (e.g., [10]).

Even when a Gibbs sampler is easy to implement, its mixing can be arbitrarily slow. This happens especially when there is a high degree of posterior dependence among the pieces of θ, such as when some pieces are highly correlated, or when the posterior density exhibits multiple modes offset "diagonally" from each other. Mixing may be improved by alternating Gibbs sampler cycles with steps of some other kind of MCMC, or by special modifications described in the following subsections.

4.2 Auxiliary variables

Gibbs sampling can be facilitated by techniques that involve sampling more than just the parameter θ. *Data augmentation* involves adding latent variables, usually as intermediaries in a hierarchical structure, that make full conditionals easier to sample. *Parameter expansion* involves creating extra dimensions in the parameter space that do not affect the Bayesian model, but allow a faster-mixing Markov chain to be constructed.

Data augmentation is natural in models that are defined using random effects. The random effects simply become latent variables to be sampled with the parameters. But it can also be used to add purely artificial latent variables designed to make full conditionals easy to sample. For example, data modeled with a location-scale t-distribution lacks any direct partial conjugacy properties. Nonetheless, a t-distribution can be represented as a scale mixture of normal distributions, with an inverse gamma distribution for the scale (variance). The scale variables then become the latent variables. Both the normal and inverse gamma distributions enjoy partial conjugacy properties that make full conditionals easy to sample. See [7], Section 12.1, for details.

Parameter expansion involves defining a redundant parameter ρ unrelated to the model itself and supplying it with an arbitrary prior density. The expanded parameterization (θ, ρ) is then reparameterized in a way specially chosen to improve Gibbs sampler performance. A basic example can be found in Section 12.1 of [7]. It is sometimes possible to use an improper prior on ρ. This leads to a Gibbs sampler that lacks a stationary distribution, but may still be able to produce valid posterior samples (see [11]).

Parameter expansion is typically used in conjunction with data augmentation, whence it is known as parameter expansion-data augmentation (PX-DA) [12].

4.3 Slice sampling

One general-purpose method to sample from an arbitrary univariate continuous density is to first sample uniformly from the bivariate (unit area) region beneath its graph and then retain only the horizontal coordinate. The uniform sampling could be performed by a simple two-step Gibbs sampler, alternating between vertical and horizontal sampling. This general approach is called *slice sampling* [13]. It can be interpreted as a special auxiliary variables method, with the vertical coordinate representing the auxiliary variable.

For a multivariate θ, slice sampling can be performed on one univariate piece at a time, as in a Gibbs sampler. Specifically, if θ_k is continuous and univariate, then the slice sampler first samples v uniformly from interval $(0, f(\theta))$, then samples θ_k uniformly from $\{\theta_k : f(\theta) > v\}$. Sampling is simplest when the latter set is an interval with easily computed endpoints, but adaptive methods are available for when this is not the case [13].

Though multivariate versions of slice sampling exist (e.g., [14]), practical implementations are often univariate and implemented as a single step within a Gibbs sampler framework, for continuous pieces that would otherwise be difficult to sample.

4.4 Metropolis-Hastings

A general approach to posterior sampling is to perform a carefully controlled random walk over the parameter space. The steps are chosen such that the resulting Markov chain has the posterior as its stationary distribution. This is accomplished by the *Metropolis-Hastings algorithm*.

In one popular version, the properties of the algorithm are determined by the choice of a random walk. The choice is arbitrary, but it is often such that each step is easy to simulate and can transition from any point in the parameter space to any other point. Let $T(\theta'|\theta)$ be its *transition kernel* for a step from θ to θ'. For example, if θ is chosen according to some continuous distribution with density $\tilde{\pi}(\theta)$, then taking one step of the random walk from θ to θ' will result in θ' having density

$$\int T(\theta'|\theta)\,\tilde{\pi}(\theta)\,d\theta \tag{6}$$

(We assume T is time-invariant, although this is not necessary, provided the time dependence does not depend on the history of the Markov chain.) The density $T(\cdot|\theta)$ defines the *proposal distribution* when the current state is θ. Values of this density (up to a constant factor that does not depend on θ) must be computable.

The transitions of the Markov chain are determined by the following algorithm: let θ^{old} be the current state of the chain. Then

1. Sample *proposal* θ' from the proposal distribution at θ^{old}.

2. Compute

$$\alpha = \frac{f(\theta')/T(\theta'|\theta^{\text{old}})}{f(\theta^{\text{old}})/T(\theta^{\text{old}}|\theta')} \tag{7}$$

3. Set the next state of the chain to be

$$\theta^{\text{new}} = \begin{cases} \theta' & \text{with probability} \min(\alpha, 1) \\ \theta^{\text{old}} & \text{otherwise} \end{cases} \tag{8}$$

Note the possibility that the next state of the chain will be identical to the previous state, even if θ is continuous under the posterior. If θ' actually becomes the next state of the chain, we say that the proposal is *accepted*. The long-run fraction of times the proposal is accepted is the *acceptance rate*.

General proof that this algorithm produces a Markov chain with the posterior as its stationary distribution can be found in, for example, [15]. Convergence properties have been extensively studied [4].

One important special case is the *Metropolis algorithm*, in which the transition kernel is symmetric: $T(\theta'|\theta) = T(\theta|\theta')$. In this case, T cancels from Eq. (7), so there is no need to compute its values. If parameter θ is continuous on an open subset of a space of real vectors, a typical example is a multivariate normal proposal distribution centered at the current value (θ^{old}). The covariance matrix is arbitrary and can be chosen to make the sampling more efficient.

Proposal distributions often admit a choice of scaling that can be tuned to improve sampling efficiency. Setting the scale too large leads to a low acceptance rate, hence slow mixing due to many repeated values. Setting the scale too small leads to a high acceptance rate, but each proposal will be close to the current value, and hence the mixing will also be slow. In some cases, theoretical results are available to guide the choice of scale. For example, for the Metropolis algorithm, research suggests that the optimal acceptance rate is about 0.44 for a one-dimensional θ and quickly falls to about 0.23 as the dimension of θ increases [16].

In addition, the shape of the proposal distribution can often be tuned. Perhaps the best shapes are ones that approximate the shape of the posterior distribution, since then proposals will tend to be in directions in which the posterior is wider. While the exact shape of the posterior may not be obvious, it may still be possible to choose a proposal that has a similar covariance structure.

The scale and shape of the proposal can be tuned in an automated manner, by making a preliminary run of the algorithm during which features of the proposal are modified adaptively to improve efficiency. This stage of *adaptation* occurs prior to burn-in: The algorithm is not a Markov chain when the proposal distribution is changed based on the sampling history, so it may not be converging to the posterior. Once adaptation is declared complete, the proposal distribution is kept fixed for burn-in and for sampling.

Although it may not be obvious, exact Gibbs sampling can actually be viewed as a special case of Metropolis-Hastings (e.g., [3]). The α turns out to always equal 1 for this situation, so no tuning is needed. Also, in a Gibbs sampler context, when a piece of θ cannot be easily simulated using conventional methods, its Gibbs step may be replaced with an easier step of Metropolis for the full conditional of that piece.

Since the posterior density is analytically available (up to a constant factor), its local properties may suggest an efficient choice of proposal distribution. For a continuous posterior, *Langevin* methods use the gradient of the log posterior density at the current point to adaptively choose the proposal distribution (e.g., [16]). This provides higher optimal acceptance rates and better scaling properties than pure Metropolis, though at the expense of more computation for each step. While this is an important improvement, modern practice has evolved even further to use more global properties of the posterior density, as detailed in the next subsection.

4.5 Hamiltonian Monte Carlo

Hamiltonian Monte Carlo (HMC), also called *hybrid* Monte Carlo, can be regarded as a special case of Metropolis-Hastings that uses a proposal involving a special set of auxiliary variables and the path of a carefully devised differential equation [17]. Computing each proposal is complicated, and perhaps expensive, but this is often compensated by achieving a high acceptance rate even when the step size is large. This results in a sequence of samples that are less dependent, and hence fewer are needed to achieve high approximation accuracy.

HMC can be applied directly to θ if the posterior is continuous and its density is continuously differentiable. Let p represent a vector of auxiliary variables having the same size as θ, but independent of θ. Specify for p an easily-sampled continuous distribution (often multivariate normal) with a continuously differentiable density proportional to $g(p)$. Define

$$H(\theta,p) = -\ln f(\theta) - \ln g(p) \tag{9}$$

Then apply the Metropolis algorithm to sample (θ,p) jointly, with proposals generated as follows:

1. Directly generate p (independently of θ).

2. Starting from (θ^{old},p), follow the path $(\theta(t),p(t))$ of the differential equation system defined by

$$\frac{d\theta_k}{dt} = \frac{\partial H}{\partial p_k} \qquad \frac{dp_k}{dt} = -\frac{\partial H}{\partial \theta_k} \tag{10}$$

   for each element θ_k of θ and corresponding element p_k of p, up to a predetermined point t_L.

3. Let $(\theta(t_L), p(t_L))$ be the new proposed value.

In the Metropolis acceptance step, we use

$$\alpha = \exp\left\{H\left(\theta^{\text{old}}, p\right) - H(\theta(t_L), p(t_L))\right\} \tag{11}$$

Actually, if the path is followed exactly, the acceptance probability will always be 1, since value of H is constant along any differential equation path [17]. The Metropolis step is needed only because, in practice, a numerical approximation is used to solve the differential equation.

To follow the differential equation numerically, we use the *leapfrog* method [17]. This method has a number of advantages over competing methods, including stability (better preservation of H) and volume preservation, which makes Metropolis valid (i.e., makes the joint transition kernel defined by this process symmetric).

If θ is not entirely continuous, HMC may still be applicable to the continuous pieces of θ, for example, when used as part of a Gibbs sampler. Also, if the posterior density is nonzero only over a certain region, HMC can be adapted for that situation. For example, it is possible to place lower and upper bounds on the elements of θ [17].

The differential equation path of an HMC proposal has a tendency to loop back on itself, making the efficiency sensitive to the length of the path (i.e., the choice of t_L). The no-U-turn sampler (NUTS) [18] is a modification of HMC designed to avoid this behavior. Essentially, it allows for adaptive choice of the leapfrog algorithm's step size and number of steps.

In theory, the computational cost of HMC scales better with the dimension of θ than does the computational cost of ordinary (random-walk) Metropolis methods. An extensive theoretical comparison can be found in [17].

5. Cluster sampling and variation

The first non-local or cluster sampling for Monte Carlo simulation for large systems is the *Swendsen-Wang* (SW) *algorithm* [19]. It was designed for the Ising and Potts models and was later generalized to other systems. The main component was the random cluster model, represented via percolation models of connecting bonds. Let us start with a spin configuration $\{\sigma\}$ and generate a percolation configuration based on the spin configuration. Next, the old spin configuration is forgotten and a new spin configuration $\{\sigma'\}$ is generated according to percolation. The rule for the process is defined in order for the detailed balance condition to be satisfied. In this way, the transition leaves the equilibrium probability invariant.

Consider a Potts model with probability distribution

$$g(\sigma) = \frac{1}{Z}\exp\left(K\sum_{<i,j>}\left(\delta_{\sigma_i,\sigma_j} - 1\right)\right) \tag{12}$$

where K is the coupling strength; the spins take on the values $1, 2, \ldots, q$, e.g. $\sigma_i = 1, 2, \ldots, q$; $\delta_{\sigma_i,\sigma_j}$ is the Kronecker delta, which equals one whenever $\sigma_i = \sigma_j$ and zero otherwise; the summation goes through nearest neighbor pairs; and Z is the partition function.

A SW Monte Carlo move is based on the following two steps: the first step transforms a Potts configuration to a bond configuration, and the second transforms back from bond to a new Potts configuration.

1. If $\sigma_i = \sigma_j$, a bond $n_{ij} = 1$ is created stochastically between neighbor sites i and j with a probability of $1 - (e)^{-K}$. Otherwise, no bond will be present and the bond variable is set to $n_{ij} = 0$.

2. Clusters are identified as sets of sites connected by bonds (otherwise isolated sites). If there is a connected path of bonds joining two sites, they are said to be in the same cluster. A new Potts value is assigned to each cluster, chosen with equal probability among 1 to q. The new Potts variable σ' is determined as the value of the cluster it belongs to.

With this approach, every state can be reached from any other state in one move with a non-zero probability. The two steps leave the probability distribution invariant and the method generates an equilibrium distribution Eq. (12).

One variation of the SW method is generalizing it to arbitrary sampling probabilities defined on graph partitions, which is achieved through considering it as a Metropolis-Hastings algorithm and computing the acceptance probability of the proposed Monte Carlo move [20]. The new inference algorithm begins by calculating graph edge weights using local image features and then is followed by two iterative steps: *Cluster Graph*: cutting the edges probability using their weights, to form connected components; *Relabel Graph*: selecting one connected component, and simultaneously flipping the partition of all its vertices in a probabilistic way. Accordingly, instead of flipping a single vertex as in Gibbs sampler, the split, merge, and re-grouping of a chunk of the graph are realized with this strategy.

The generalized cluster sampling implements ergodic and reversible Markov chain jumps on graph partitions. It is applicable to arbitrary posterior probabilities or energy functions in the space of graphs. Examples in image analysis (e.g., image segmentation) demonstrate that the cluster Monte Carlo is more efficient than the classical Gibbs sampler and performs better than the graph cuts and belief propagation.

6. Software implementation

In the statistics community, the first development of practical general-purpose software for MCMC was the BUGS (Bayesian inference using Gibbs sampling) project, starting in 1989. The original implementation, designed for the Windows operating system, was WinBUGS, which included a graphical interface. When development of WinBUGS ended, the OpenBUGS project was created as a successor. This software uses a special model specification language, the "BUGS language," that is remarkably flexible. Usually, the analyst only needs to specify the model in the BUGS language and then leave the construction of appropriate samplers to the software. The basic structure is a Gibbs sampler, but the pieces may be sampled using specialized methods.

Inspired by BUGS, a parallel effort called JAGS (Just another Gibbs sampler) was developed. Like BUGS, it is based on Gibbs sampling and, in principle, requires the analyst to specify only a model (written in a variant of the BUGS language), leaving the construction of samplers to an automated engine. It tends to be faster than OpenBUGS, is more actively developed, and features better integration with the R language. It also incorporates efficient slice samplers in some of its steps. JAGS is entirely open-source and has versions for many operating systems.

PyMC's development was an effort to generalize the process of building Metropolis-Hastings samplers, making MCMC more accessible to non-statisticians. It is now a Python package helping users define stochastic models and construct

Bayesian posterior samples. A large number of problems are suitable for PyMC due to its flexibility and extensibility. Key features include ability to fit Bayesian statistical models via MCMC and other algorithms; a large set of well-documented statistical distributions; a module for modeling Gaussian processes; sampling loops can be manipulated manually, etc.

A more recently introduced tool (since 2012) is the language Stan, which remains under active development (as of this writing). Stan allows model specification, but in an inherently more flexible way than BUGS or its variants. Software for compiling Stan includes the option for MCMC using HMC and NUTS. It therefore tends to produce more nearly independent samples than software based on Gibbs sampling. (There are also options for inference not based on sampling, such as variational methods.) The Stan software integrates with R, Python, MATLAB, Julia, and Stata.

7. Conclusion

This chapter has merely touched upon the important concepts and methods of modern MCMC. Routine-use software automating the construction of samplers is also introduced. There are many good references that provide more detailed theoretical or practical treatment and further extensions, based on which future research can be developed.

Author details

Michelle Y. Wang[1,2,3*] and Trevor Park[1]

1 Department of Statistics, University of Illinois at Urbana-Champaign, USA

2 Department of Psychology, University of Illinois at Urbana-Champaign, USA

3 Department of Bioengineering, University of Illinois at Urbana-Champaign, USA

*Address all correspondence to: ymw@illinois.edu

References

[1] Tierney L, Kadane JB. Accurate approximations for posterior moments and marginal densities. Journal of the American Statistical Association. 1986; **81**(393):82-86

[2] Jordan MI, Ghahramani Z, Jaakkola TS, Saul LK. An introduction to variational methods for graphical models. Machine Learning. 1999;**37**: 183-233

[3] Geyer CJ. Introduction to Markov chain Monte Carlo. In: Brooks S, Gelman A, Jones GL, Meng X-L, editors. Handbook of Markov Chain Monte Carlo. Boca Raton: CRC Press; 2011. pp. 3-48

[4] Tierney L. Markov chains for exploring posterior distributions. Ann. Statist. 1994;**22**(4):1701-1728

[5] Gelman A, Shirley K. Inference from simulations and monitoring convergence. In: Brooks S, Gelman A, Jones GL, Meng X-L, editors. Handbook of Markov Chain Monte Carlo. Boca Raton: CRC Press; 2011. pp. 163-174

[6] Flegal JM, Jones GL. Implementing MCMC: Estimating with confidence. In: Brooks S, Gelman A, Jones GL, Meng X-L, editors. Handbook of Markov Chain Monte Carlo. Boca Raton: CRC Press; 2011. pp. 175-197

[7] Gelman A, Carlin JB, Stern HS, Dunson DB, Vehtari A, Rubin DB. Bayesian Data Analysis. 3rd ed. Boca Raton: CRC Press; 2013. p. 667

[8] Roberts GO, Sahu SK. Updating schemes, correlation structure, blocking and parameterization for the Gibbs sampler. Journal of the Royal Statistical Society: Series B: Methodological. 1997; **59**(2):291-317

[9] Van Dyk DA, Park T. Partially collapsed Gibbs samplers: Theory and methods. Journal of the American Statistical Association. 2008;**103**(482): 790-796

[10] Van Dyk DA, Jiao X. Metropolis-Hastings within partially collapsed Gibbs samplers. Journal of Computational and Graphical Statistics. 2015;**24**(2):301-327

[11] Hobert JP. Stability relationships among the Gibbs sampler and its subchains. Journal of Computational and Graphical Statistics. 2001;**10**(2): 185-205

[12] Liu JS, Wu YN. Parameter expansion for data augmentation. Journal of the American Statistical Association. 1999;**94**:1264-1274

[13] Neal RM. Slice sampling. Ann. Statist. 2003;**31**(3):705-741

[14] Damien P, Wakefield J, Walker S. Gibbs sampling for Bayesian non-conjugate and hierarchical models by using auxiliary variables. Journal of the Royal Statistical Society, Series B: Statistical Methodology. 1999;**61**(2): 331-344

[15] Tierney L. A note on Metropolis-Hastings kernels for general state spaces. Ann. Appl. Probab. 1998;**8**(1):1-9

[16] Roberts GO, Rosenthal JS. Optimal scaling for various Metropolis-Hastings algorithms. Statistical Science. 2001; **16**(4):351-367

[17] Neal RM. MCMC using Hamiltonial dynamics. In: Brooks S, Gelman A, Jones GL, Meng X-L, editors. Handbook of Markov Chain Monte Carlo. Boca Raton: Chapman & Hall; 2011. pp. 113-162

[18] Hoffman MD, Gelman A. The No-U-turn sampler: Adaptively setting path

lengths in Hamiltonian Monte Carlo. Journal of Machine Learning Research. 2014;**15**:1593-1623

[19] Swendsen RH, Wang JS. Nonuniversal critical dynamics in Monte Carlo simulations. Physical Review Letters. 1987;**58**(2):86-88

[20] Barbu A, Zhu SC. Generalizing Swendsen-Wang to sampling arbitrary posterior probabilities. IEEE Transactions on Pattern Analysis and Machine Intelligence. 2005;**27**(8): 1239-1253

Chapter 3

A Review on the Exact Monte Carlo Simulation

Hongsheng Dai

Abstract

Perfect Monte Carlo sampling refers to sampling random realizations exactly from the target distributions (without any statistical error). Although many different methods have been developed and various applications have been implemented in the area of perfect Monte Carlo sampling, it is mostly referred by researchers to coupling from the past (CFTP) which can correct the statistical errors for the Monte Carlo samples generated by Markov chain Monte Carlo (MCMC) algorithms. This paper provides a brief review on the recent developments and applications in CFTP and other perfect Monte Carlo sampling methods.

Keywords: coupling from the past, diffusion, Monte Carlo, perfect sampling

1. Introduction

In the past 30 years, substantial progress has been made in popularizing Bayesian methods for the statistical analysis of complex data sets. An important driving force has been the availability of different types of Bayesian computational methods, such as Markov chain Monte Carlo (MCMC), sequential Monte Carlo (SMC), approximate Bayesian computation (ABC) and so on. For many practical examples, these computational methods can provide rapid and reliable approximations to the posterior distributions for unknown parameters.

The basic idea that lies behind these methods is to obtain Monte Carlo samples from the distribution of interest, in particular the posterior distribution. In Bayesian analysis of complex statistical models, the calculation of posterior normalizing constants and the evaluation of posterior estimates are typically infeasible either analytically or by numerical quadrature. Monte Carlo simulation provides an alternative. One of the most popular Bayesian computational methods is MCMC, which is based on the idea of constructing a Markov chain with the desired distribution as its stationary distribution.

By running a Markov chain, MCMC methods generate statistically dependent and approximate realizations from the limiting (target) distribution. A potential weakness of these methods is that the simulated trajectory of a Markov chain will depend on its initial state. A common practical recommendation is to ignore the early stages, the so-called *burn-in* phase, before collecting realizations of the state of the chain. How to choose the length of the *burn-in* phase is an active research area. Many methods have been proposed for *convergence diagnostics*; [1] gave a comparative review. Rigorous application of diagnostic methods requires either substantial empirical analysis of the Markov chain or complex mathematical analysis.

In practice, judgments about convergence are often made by visual inspection of the realized chain or the application of simple rules of thumb.

Concerns about the *quality* of the sampled realizations of the simulated Markov chains have motivated the search for Monte Carlo methods that can be guaranteed to provide samples from the target distribution. This is usually referred to as *perfect sampling* or *exact sampling*. In some cases, for example, the multivariate normal, perfect samples are readily available. For more complicated distributions, perfect sampling can be achieved, in principle, by the *rejection method*. This involves sampling from a density that bounds a suitable multiple of the target density, followed by acceptance or rejection of the sampled value. The difficulty here is in finding a bounding density that is amenable to rapid sampling while at the same time providing sample values that are accepted with high probability. In general this is a challenging task, although efficient rejection sampling methods have been developed for the special class of log-concave densities; see, for example, [2, 3].

A surprising breakthrough in the search for perfect sampling methods was made by [4]. The method is known as *coupling from the past* (CFTP). This is a sophisticated MCMC-based algorithm that produces realizations exactly from the target distribution. CFTP transfers the difficulty of running the Markov chain for extensive periods (to ensure convergence) to the difficulty of establishing whether a potentially large number of coupled Markov chains have coalesced. An efficient CFTP algorithm depends on finding an appropriate Markov chain construction that will ensure fast coalescence. There have been a few further novel theoretical developments following the breakthrough of CFTP, including [2, 5, 6]. Since then, perfect sampling methods have attracted great attention in various Bayesian computational problems and applied probability areas.

Apart from coupling from the past, many other perfect sampling methods were proposed for specific problems, for example, perfect sampling for random spanning trees [7, 8] and path-space rejection sampler for diffusion processes [9–11]. Very recently, a type of divide-and-conquer method has been developed in [12, 13]. These methods use the technique for the exact simulation of diffusions and samples from simple sub-densities to obtain perfect samples from the target distribution.

Perfect samples are useful in Bayesian applications either as a complete replacement for MCMC-generated values or as a source of initial values that will guarantee that a conventional MCMC algorithm runs in equilibrium. Perfect samples can also be used as a means of quality control in judging a proposed MCMC implementation when there are questions about the speed of convergence of the MCMC algorithm or whether the chain is capable of exploring all parts of the sample space. Of course, when perfect samples can be obtained speedily, they will be preferred to MCMC values, since they eliminate such doubts. In addition, sophisticated perfect sampling methodology often motivates efficient approximate algorithms and computational techniques. For example, [14] uses regenerated Markov chains to obtain simple standard error estimates for importance sampling under MCMC context. The condition required there will allow us to carry out perfect sampling via multigamma coupling approach [15].

This paper will present a brief review for perfect Monte Carlo sampling and explain the advantages and drawbacks of different types of methods. Section 2 will present rejection sampling techniques, and then CFTP will be covered in Section 3. Recent divide-and-conquer methods will be reviewed in Section 4. The paper ends with a discussion in Section 5.

2. Rejection sampling techniques

Rejection sampling, also known as 'acceptance-rejection sampling', generates realizations from a target probability density function $f(x)$ by using a hat function

$Mg(x)$, where $f(x) \leq Mg(x)$ and $g(x)$ are a probability density function from which samples can be readily simulated. The basic rejection sampling algorithm is as follows:

Algorithm 2.1 (Basic rejection sampling)

01 Sample x from $g(x)$ and U from Unif$[0, 1]$.
02 If $U \leq \frac{f(x)}{Mg(x)}$,
03   accept x as a realisation of $f(x)$ and stop;
04 else
05   reject the value of x and go back to step 01.

Many other perfect sampling methods are actually equivalent to rejection sampling. For example, ratio-of-uniform (RoU) method [16] may have to be implemented via a rejection sampling approach.

The efficiency of rejection sampling depends on the acceptance probability, which is $1/M$. To perform rejection sampling efficiently, it is very important to find hat functions which provides higher acceptance probabilities. In other words, we shall choose M as small as possible [17]; however for many complicated problems, there is no easy way to find M small enough to guarantee high acceptance probability. A number of sophisticated rejection sampling methods have been suggested for dealing with complex Bayesian posterior densities, which we discuss below.

2.1 Log-concave densities

A function $h(x)$ is called log-concave if

$$\log h(\lambda x + (1-\lambda)y) \geq \lambda \log h(x) + (1-\lambda) \log h(y),$$

for all x, y and $\lambda \in [0, 1]$. For the special class of log-concave densities, Gilks and wild [3] developed the adaptive rejection sampling (ARS) method. The method constructs an envelope function for the logarithm of the target density, $f(x)$, by using tangents to $\log f(x)$ at an increasing number, n, of points. The envelope function $u_n(x)$ is the piecewise linear *upper hull* formed from the tangents. Note that, the envelope can be easily constructed due to the concavity of $\log f(x)$. The method also constructs a squeeze function $l_n(x)$ which is formed from the chords of the tangent points. The sampling steps of the ARS algorithm are as follows.

Algorithm 2.2 (Adaptive rejection sampling).

Outputs a stream of perfect samples from $f(x)$.
01 Initialise $u_n(x)$ and $l_n(x)$ by using tangents at several points.
02 Sample x^* from density $\propto \exp(u_n(x))$ and $U \sim$ Unif$(0, 1)$
03 If $U \leq \exp(l_n(x^*) - u_n(x^*))$, Output x^*;
04 else if $U \leq f(x^*)/\exp(u_n(x^*))$,
05   Output x^*; Update (u_n, l_n) to (u_{n+1}, l_{n+1}) using x^*;
06 Goto 02

The ARS algorithm is adaptive and the sampling density is modified whenever $f(x^*)$ is evaluated. In this way, the method becomes more efficient as the sampling continues. Leydold [18] extends ARS to log-concave multivariate densities.

Leydold's algorithm is based on decomposing the domain of the density into cones and then computing tangent hyperplanes for these cones. Generic computer code for sampling from a multivariate log-concave density is available on the author's website; it is only necessary to code a subroutine for the target density. Leydold's algorithm works well for simple low-dimensional densities. The drawback of ARS algorithm is that it only works for log-concave densities, which is a very small class of posteriors in practice. Also computationally it is very challenging to implement ARS algorithm for high-dimensional densities [19].

2.2 Fill's rejection sampling algorithm

We consider a discrete Markov chain with transition probability $\mathbf{P}(x,y)$ and stationary distribution $\pi(x), x \in S$. Let $\tilde{\mathbf{P}}(x, y) = \pi(y)\mathbf{P}(y,x)/\pi(x)$ be the transition probability for the time-reversed chain. Suppose that there is a partial ordering on the states S, denoted by $x \preccurlyeq y$. Let $\hat{0}$ and $\hat{1}$ be unique extremal points of the partial ordering.

To demonstrate the algorithm given by [2], we will assume that there are update functions ϕ and $\tilde{\phi}$ both mapping $S \times [0,1]$ to S such that

$$\mathbf{P}(x,y) = P(\phi(x,U) = y), \tag{1}$$

$$\tilde{\mathbf{P}}(x,y) = P(\tilde{\phi}(x,U) = y), \tag{2}$$

where $U \sim \text{Unif}[0,1]$ and

$$x \preccurlyeq y \Rightarrow \tilde{\phi}(x,u) \preccurlyeq \tilde{\phi}(y,u) \quad a.e. \quad u \in [0,1].$$

The algorithm is as follows:

Algorithm 2.3 (Fill's algorithm)

---

1. Choose an integer $t > 0$. Fix $x_0 = \hat{0}$ and $y_0 = \hat{1}$.
2. Generate $x_k = \phi(x_{k-1}, U_k), k = 1, \cdots, t$, where $\{U_k\}$ are i.i.d. Unif$[0,1]$.
3. Generate $\tilde{U}_k$ from the conditional distribution of U given $\phi(x_{t-k+1}, U) = x_{t-k}, k = 1, \cdots, t$.
4. Generate $y_k = \tilde{\phi}(y_{k-1}, \tilde{U}_k), k = 1, \cdots, t$.
5. If $y_t = x_0$ then accept x_t; else double t and repeat from step 2.

In Algorithm 2.3 (step 2) $z := x_t$ is sampled from $\mathbf{P}^t(\hat{0}, \cdot)$. If we find an upper bound M for $\pi(z)/\mathbf{P}^t(\hat{0}, z)$, then we can use rejection sampling. Fill [2] finds a bounding constant given by $M = \pi(\hat{0})/\tilde{\mathbf{P}}^t(\hat{1}, \hat{0})$ and proves that steps from 3 to 5 of Algorithm 2.3 are to accept z with probability $\frac{\pi(z)}{M\mathbf{P}^t(\hat{0}, z)}$. The output of this algorithm is indeed a perfect sample from π.

From Algorithm 2.3, we can see that rejection sampling can still be possible, even if we do not have a closed form of the hat function. The first limitation of Algorithm 2.3 is that it works only if the time-reversed chain is monotone, but [5] has extended the algorithm theoretically for general chains. The second limitation is that step 3 of Algorithm 2.3 is difficult to perform [2]. For these reasons, Fill's algorithm is not practical for realistic problems.

DOI: http://dx.doi.org/10.5772/intechopen.88619

3. Coupling from the past

Coupling from the past was introduced in the landmark paper of [4] which showed how to provide perfect samples from the limiting distribution of a Markov chain.

3.1 Basic CFTP algorithms

Let $\{X_t\}$ be an ergodic Markov chain with state space $\mathcal{X} = \{1, \cdots, n\}$, where the probability of going from i to j is p_{ij} and the stationary distribution is π. Suppose we design an updating function $\phi(\cdot, U)$, which satisfies $P[\phi(i, U) = j] = p_{ij}$, where ϕ is a deterministic function and U is a random variable. To simulate the next state Y of the Markov chain, currently in state i, we draw a random variable U and let $Y = \phi(i, U)$.

Let $f_t(i) = \phi(i, U_t)$, and define the composition

$$F_{t_1}^{t_2} = f_{t_2-1} \circ f_{t_2-2} \circ \cdots \circ f_{t_1+1} \circ f_{t_1}, \tag{3}$$

for $t_1 < t_2$.

Proposition 3.1 [4] Suppose there exists a time $t = -T$, the backward coupling time, such that chains starting from any state in $\mathcal{X} = \{1, \cdots, n\}$, at time $t = -T$, and with the same sequence $\{U_t, t = -T, \cdots, -1\}$, arrive at the same state X_0^*. Then it must follow that X_0^* comes from π.

This proposition is easy to prove. If we run an ergodic Markov chain from time $t = -\infty$ and with the sequence $\{U_t, t = -T, \cdots, -1\}$ after $-T$, the Markov chain will arrive at X_0^*. Then X_0^* comes exactly from π since it is collected at time 0 and the Markov chain started from $-\infty$. The algorithm is as follows:

Algorithm 3.1 (Basic CFTP)

$t = 0$
repeat
  $t = t - 1$
  generate U_t
  $F_t^0 = F_{t+1}^0 \circ \phi(\cdot, U_t)$
until $F_t^0(\cdot)$ is a constant
return $F_t^0(\cdot)$

Propp and Wilson [4] also proved that this algorithm is certain to terminate. The idea of simulating from the past is important. Note that if we collect $F_0^T(\cdot)$ as the result, where T is the smallest value that makes $F_0^T(\cdot)$ a constant, we will get a biased sample. This is because T is a stopping time, which is not independent of the Markov chain.

3.2 Read-once CFTP

The basic CFTP algorithm needs to restart the Markov chains at some points in the past if they have not coalesced by time 0. We must use the same random sequence $\{U_t\}_{-\infty}^{-1}$ when we restart the Markov chains. In Monte Carlo simulations, we usually use *pseudo*random number generators, which are deterministic algorithms. So if we give the same random seed, we will get the same random sequence. This means that the same sequence $\{U_t\}$ can be regenerated in each coupling procedure.

If we can run the Markov chain forward starting at 0 and collect a perfect sample in the future, we will not have to regenerate $\{U_t\}$. Wilson [20] developed a read-once CFTP method to implement the forward coupling idea. A simple example is provided by [21]. In fact, the multigamma coupler in [15] can be implemented via the more efficient read-once CFTP algorithm.

3.3 Improvement of CFTP algorithms

Propp and Wilson [4] showed that the computational cost of the algorithm can be reduced if there is a partial order for the state space $\mathcal{X}$ that is preserved by the update function ϕ, i.e. if $x \leq y$ then $\phi(x, U) \leq \phi(y, U)$. Their procedure is outlined in Algorithm 3.2, whereas before $\hat{0}$ and $\hat{1}$ are the unique extremals. Note that their algorithm needs a monotone update function ϕ for the Markov chain, while Algorithm 2.3 requires a monotone update function $\tilde{\phi}$ for the time-reversed chain.

Algorithm 3.2 (Monotone CFTP)

$T = 1$
Repeat
    upper $= \hat{1}$
    lower $= \hat{0}$
    for $t = -T$ to $t = -1$
        upper $= \phi_t(\text{upper}, U_t)$
        lower $= \phi_t(\text{lower}, U_t)$
    $T = 2T$
until upper = lower
return upper

Algorithm 3.2 is much simpler than Algorithm 3.1, since only two chains have to be run at the same time, but the requirement of monotonicity is very restrictive. Markov chains with transitions given by independent Metropolis-Hastings and perfect slice sampling have been shown to have this property, by [22, 23], respectively. However [23, 24] have also noticed that such independent M-H CFTP is equivalent to simple rejection sampler.

In general it is hard to code perfect slice samplers correctly. For example, Hörmann and Leydold [25] have pointed out that the perfect slice samplers in [26, 27] are incorrect. The challenge of monotone CFTP is usually to construct the detailed updating function with a guarantee of preserving the partial order.

Finding a partial order preserved by the Markov chains is a non-trivial task in many cases. An alternative improvement is to use CFTP with bounding chains, such as that in [28, 29]. If the bounding chains, which bound all the Markov chains, coalesce, then all Markov chains coalesce. Thus if only a few bounding chains are required, the efficiency of the CFTP algorithm can be improved significantly. Sometimes, it may be impossible to define an explicit bounding chain (the maximum of the state space may be infinity, and the upper bound chain cannot start from infinity), but it is possible to use a dominated process to bound all Markov chains [30].

3.4 Applications and challenges

Although CFTP is extremely challenging to be implemented for many practical problems, it did find a few applications in certain discrete state space problems, for example, the Ising model [4]. Also [31] applied CFTP to ancestral selection graph to simulate samples from population genetic models. Refs. [32, 33] applied CFTP to a class of fork-join type queuing system problems. Connor and Kendal [34] applied CFTP for the perfect simulation of M/G/c queues. CFTP also finds its application in signal processing [35].

CFTP for continuous state space Markov chains is very challenging, since a random map from an interval to a finite number of points is required. In recent years, many methods have been developed for unbounded continuous state space Markov chains, such as perfect slice sampler in [23], multigamma coupler and the bounded M-H coupler in [15, 24]. Wilson [36] developed a layered multi-shift coupling, which shifts states in an interval to a finite number of points. However, none of these methods can solve any practical problems.

4. Recent advances in perfect sampling

Recently, a new type of perfect Monte Carlo sampling method based on the decomposition of the target density f, as $f(\cdot) = g_1(\cdot)g_2(\cdot)$, was proposed in [12], where g_1 and g_2 are also (proportional to) proper density functions. Note that here g_1 and g_2 are continuous density functions which are easy to simulate from. Suppose that q-dimensional vector values $\boldsymbol{x}_1$ and $\boldsymbol{x}_2$ are independently drawn from g_1 and g_2, respectively. If the two independent samples are equal, i.e. $\boldsymbol{x}_1 = \boldsymbol{x}_2 = \boldsymbol{y}$ then we have $\boldsymbol{y}$ must be from $f(\cdot) \propto g_1(\cdot)g_2(\cdot)$. Note that such a naive approach may be practical for discrete random variables with low-dimensional state space, but for continuous random variables, it is impossible since $\mathrm{P}(\boldsymbol{x}_1 = \boldsymbol{x}_2) = 0$. Dai [12] proposed a novel approach to deal with this, which is explained in the following subsection.

4.1 Perfect distributed Monte Carlo without using hat functions

First we introduce the following notations related to the logarithm of the sub-densities:

$$\boldsymbol{\alpha}(\boldsymbol{x}) = \left(\alpha^{(1)}, \cdots, \alpha^{(q)}\right)^{tr}(\boldsymbol{x}) = \nabla \log g_1(\boldsymbol{x}) \tag{4}$$

where ∇ is the partial derivative operator for each component of $\boldsymbol{x}$. Then we consider a q-dimensional diffusion process $\boldsymbol{X}_t\left(\vec{\boldsymbol{\omega}}\right), t \in [0, T]$ $(T < \infty)$, defined on the q-dimensional continuous function space $\boldsymbol{\Omega}$, given by:

$$d\boldsymbol{X}_t = \boldsymbol{\alpha}(\boldsymbol{X}_t)dt + d\boldsymbol{B}_t, \tag{5}$$

where $\boldsymbol{B}_t\left(\vec{\boldsymbol{\omega}}\right) = \omega_t$ is a Brownian motion and $\vec{\boldsymbol{\omega}} = \{\omega_t, t \in [0, T]\}$ is a typical element of $\boldsymbol{\Omega}$. Let $\mathbb{W}$ be the probability measure for a Brownian motion with the initial probability distribution $\boldsymbol{B}_0 = \boldsymbol{w}_0 \sim f_1(\cdot) = g_1^2(\cdot)$.

Clearly $\boldsymbol{X}_t$ has the invariant distribution $f_1(\boldsymbol{x})$ (using the Langevin diffusion results [37]). Let $\mathbb{Q}$ be the probability law induced by $\boldsymbol{X}_t, t \in [0, T]$, with $\boldsymbol{X}_0 = \omega_0 \sim f_1(\cdot)$, i.e. under $\mathbb{Q}$ we have $\boldsymbol{X}_t \sim f_1(\boldsymbol{x})$ for any $t \in [0, T]$.

The idea in [12] is to use a *biased* diffusion process $\overline{\mathbf{X}} = \left\{\overline{\boldsymbol{X}}_t; 0 \le t \le T\right\}$ to simulate from the target function f. It is defined as follows.

Definition 4.1 (Biased Langevin diffusions) The joint density for the pair $\left(\overline{\boldsymbol{X}}_0, \overline{\boldsymbol{X}}_T\right)$ (the starting and ending points of the biased diffusion process), evaluated at point $(\boldsymbol{x}, \boldsymbol{y})$, is $f_1(\boldsymbol{x})\mathbf{t}^*(\boldsymbol{y}|\boldsymbol{x})f_2(\boldsymbol{y})$, where $\mathbf{t}^*(\boldsymbol{y}|\boldsymbol{x})$ is the transition density for the diffusion process X defined in Eq. (6) from $\boldsymbol{X}_0 = \boldsymbol{x}$ to $\boldsymbol{X}_T = \boldsymbol{y}$ and $f_2(\boldsymbol{y}) = g_2(\boldsymbol{y})/g_1(\boldsymbol{y})$.

Given $\left(\overline{\boldsymbol{X}}_0, \overline{\boldsymbol{X}}_T\right)$ the process $\left\{\overline{\boldsymbol{X}}_t, 0 < t < T\right\}$ is given by the diffusion bridge driven by Eq. (6).

The marginal distribution for $\overline{\boldsymbol{X}}_T$ is $f(\boldsymbol{y})$. Therefore, to draw a sample from the target distribution $f(\boldsymbol{x})$, we need to simulate a process $\overline{\boldsymbol{X}}_t, t \in [0, T]$ from $\overline{\mathbb{Q}}$ (the law induced by $\overline{\boldsymbol{X}}$) and then $\overline{X}_T \sim f(\boldsymbol{x})$.

Simulation from $\overline{\mathbb{Q}}$ can be done via rejection sampling. We can use a *biased Brownian motion* $\left\{\overline{\boldsymbol{B}}_t; 0 \le t \le T\right\}$ as the proposal diffusion:

Definition 4.2 (Biased Brownian motion) The starting and ending points $\left(\overline{\boldsymbol{B}}_0, \overline{\boldsymbol{B}}_T\right)$ follow a distribution with a density $h(\boldsymbol{x}, \boldsymbol{y})$, and $\left\{\overline{\boldsymbol{B}}_t; 0 < t < T\right\}$ is a Brownian bridge given $\left(\overline{\boldsymbol{B}}_0, \overline{\boldsymbol{B}}_T\right)$.

Under certain mild conditions, Dai [12] proved the following lemma.

Lemma 4.1 Let $\mathbb{Z}$ be the probability law induced by $\left\{\overline{\boldsymbol{B}}_t; 0 \le t \le T\right\}$. By letting

$$h(\boldsymbol{\omega}_0, \boldsymbol{\omega}_T) = g_2(\boldsymbol{\omega}_T)g_1(\boldsymbol{\omega}_0)\frac{1}{\sqrt{2\pi T}}e^{-\frac{\|\omega_T - \omega_0\|^2}{2T}} \tag{6}$$

we have

$$\frac{d\overline{\mathbb{Q}}}{d\mathbb{Z}}\left(\vec{\omega}\right) \propto \exp\left[-\frac{1}{2}\int_0^T \left(\|\boldsymbol{\alpha}\|^2 + \operatorname{div}\boldsymbol{\alpha}\right)(\omega_t)dt\right] \tag{7}$$

where div is the divergence operator.

Condition 4.1 There exists $l > -\infty$ such that

$$\frac{1}{2}\left(\|\boldsymbol{\alpha}\|^2 + \mathbf{div}\boldsymbol{\alpha}\right)(\boldsymbol{x}) - l \ge 0. \tag{8}$$

Under Condition 4.1 the ratio (8) can be rewritten as

$$\frac{d\overline{\mathbb{Q}}}{d\mathbb{Z}}\left(\vec{\omega}\right) \propto \exp\left[-\int_0^T \left(\frac{1}{2}\left(\|\boldsymbol{\alpha}\|^2 + \mathbf{div}\boldsymbol{\alpha}\right)(\omega_t) - l\right)dt\right], \tag{9}$$

which has a value no more than 1.

Therefore we can use rejection sampling to simulate from $\overline{\mathbb{Q}}$, with proposal measure $\mathbb{Z}$. This acceptance probability (10) can be dealt with using similar methods as that in [9, 11]. The algorithm is presented below; see [12, 38] for more details.

Algorithm 4.1 (Simple distributed sampler)	
Simulate $(\boldsymbol{\omega}_0, \boldsymbol{\omega}_T)$ from density h	01
Simulate the biased Brownian bridge $\left(\overline{\boldsymbol{B}}_t, t \in (0, T)\right)$	02
Accept $\boldsymbol{\omega}_T$ as a sample from f, with probability (6); If rejected, go back to step 01.	03

Such a method is a rejection sampling algorithm, but it does not require finding a hat function to bound the target density, which is usually the main challenge of the traditional rejection sampling for complicated target densities. The above algorithm uses g_2 as the proposal density function, which does not have to bound the target f. However, it requires a bound for the derivatives of the logarithm of the sub-densities (see Condition 4.1). This is usually easier to get in practice, since the logarithm of the posterior is usually easy to deal with. Also [12] noted that we should choose sub-densities g_1 and g_2 as similar as possible, in order to achieve high acceptance probability.

Dai [12] focused on the simple decomposition of $f = g_1 g_2$, although it mentioned that for more general decomposition of $f = g_1 g_2 \cdots g_C$, a recursive method can be used. Unfortunately, a naive recursive method is very inefficient. A more sophisticated method is introduced in the following section.

4.2 Monte Carlo fusion for distributed analysis

A more efficient and sophisticated methods were proposed recently in [13], named as Monte Carlo fusion. Suppose that we consider

$$f(\boldsymbol{x}) \propto g_1(\boldsymbol{x}) \cdots g_C(\boldsymbol{x}), \tag{10}$$

where each $g_c(\boldsymbol{x})$ ($c \in \{1, \ldots, C\}$) is a density (up to a multiplicative constant). Here C denotes the number of parallel computing cores available in big data problems, and each $g_c(\boldsymbol{x})$ means the sub-posterior density based on a subset of the big data. In group decision problems, C means the number of different decisions which should be combined and $g_c(\boldsymbol{x})$ stands for the decision from each group member.

Dai et al. [13] considered simulating from the following density on extended space,

$$g\left(\boldsymbol{x}^{(1)}, \ldots, \boldsymbol{x}^{(C)}, \boldsymbol{y}\right) = \prod_{c=1}^{C} \left[g_c^2\left(\boldsymbol{x}^{(c)}\right) \cdot p_c\left(\boldsymbol{y} \,|\, \boldsymbol{x}^{(c)}\right) \cdot \frac{1}{g_c(\boldsymbol{y})} \right], \tag{11}$$

which admits the marginal density f for $\boldsymbol{y}$. Here $p_c\left(\boldsymbol{y} \,|\, \boldsymbol{x}^{(c)}\right)$ is the transition density from $\boldsymbol{x}^{(c)}$ to $\boldsymbol{y}$ for the Langevin diffusion defined in Eq. (6) associated with each sub-density g_c.

Dai et al. [13] considered a rejection sampling approach with proposal density proportional to the function

$$h\left(\boldsymbol{x}^{(1)}, \ldots, \boldsymbol{x}^{(C)}, \boldsymbol{y}\right) = \prod_{c=1}^{C} \left[g_c\left(\boldsymbol{x}^{(c)}\right) \right] \cdot \exp\left(-\frac{C \cdot \|\boldsymbol{y} - \overline{\boldsymbol{x}}\|^2}{2T} \right), \tag{12}$$

where $\overline{\boldsymbol{x}} = C^{-1} \sum_{c=1}^{C} \boldsymbol{x}^{(c)}$ and T is an arbitrary positive constant.

Simulation from the proposal h can be achieved directly. In particular, $\boldsymbol{x}^{(1)}, \ldots \boldsymbol{x}^{(C)}$ are first drawn independently from $g_1, \ldots, g_C$, respectively, and then $\boldsymbol{y}$ is simply a Gaussian random variable centred on $\overline{\boldsymbol{x}}$. This is a distributed analysis or divide-and-conquer approach. Detailed acceptance probabilities and rejection sampling algorithms can be found in [13].

The above fusion approach arises in modern statistical methodologies for 'big data'. A full dataset will be artificially split into a large number of smaller data sets, and inference is then conducted on each smaller data set and combined (see, for instance, [39–46]). The advantage for such an approach is that inference on each small data set can be conducted in parallel. Then the heavy computational cost of algorithms such as MCMC will not be a concern. Traditional methods suffer from the weakness that the fusion of the separately conducted inferences is inexact. However, the Monte Carlo fusion in [13] is an exact simulation algorithm and does not have any approximation weakness.

The above fusion approach also arises in a number of other settings, where distributed analysis came naturally. For example, in signal processing, distributed multi-sensor may be used for network fusion systems. Fusion approach arises naturally to combine results from different sensors [47].

5. Conclusion

Although perfect simulation usually refers to correcting the statistical errors for the samples drawn via MCMC, it actually covers a much wider area beyond CFTP. In fact for certain applications, it is often possible to construct other types of perfect sampling methods which are much more efficient than CFTP. For example, for the exact simulation of the posterior of simple mixture models, the geometric-arithmetic mean (GAM) method in [19] is much more efficient than CFTP in [48]. Details of GAM method is provided in Appendix. Also the random walk construction for exact simulation for random spanning trees [7] is much more efficient than the CFTP version.

Bayesian computational algorithms keep evolving, in particular under the current big data era. Although almost all newly developed algorithms are approximate simulation algorithms, perfect sampling is still one of the key wheel-driven forces for new Bayesian computational algorithms, and they usually can quickly motivate new class of 'mainstream' algorithms. More focus should be given to methods beyond CFTP, for example, the fusion type of algorithms.

The Monte Carlo fusion method has the potential to be used in many Bayesian big data applications. For example, for large car accident data, the response variable is usually a categorical variable representing the seriousness of the accident, and generalized linear regression model is often used. Under a Bayesian framework, we may estimate the posterior distribution for the regression parameters via such a fusion approach. Then the posterior mean, the posterior median, or other characteristics of the posterior distribution can be estimated using the Monte Carlo samples. Also such an algorithm is perfect sampling algorithm, and no convergence justification is needed, since it always provided realizations exactly from the target distribution.

Appendix

A. Geometric-arithmetic mean method for simple mixture model

Observations from a simple mixture model are assumed to be either discrete or continuous. The density function of an individual observation y has the form

$$f(y;\boldsymbol{p}) = \sum_{k=1}^{K} p_k f_k(y), \quad \text{where} \quad \sum_{k=1}^{K} p_k = 1, \quad \text{and} \quad p_k > 0, k = 1, \ldots, K. \qquad (13)$$

We assume that the component weights $\boldsymbol{p} = (p_1, \ldots, p_K)$ are unknown parameters and the number of components, K, and the component densities, f_k, are all known. We focus on the perfect sampling from the posterior distribution of $\mathbf{p}$.

Suppose that we have N observations, $y_1, \ldots, y_N$. Let $L_{nk} = f_k(y_n)$ and assume that the prior distribution of $\mathbf{p}$ is Dirichlet:

$$\pi_0(\mathbf{p}) \propto \prod_{k=1}^{K} p_k^{\alpha_k - 1}, \quad \alpha_k > 0, k = 1, \ldots, K. \tag{14}$$

Then the posterior distribution is given by

$$f(\mathbf{p}|y) \propto \pi_0(\mathbf{p}) \prod_{n=1}^{N} \left(\sum_{k=1}^{K} p_k L_{nk} \right) I_{\mathcal{X}}(\mathbf{p}), \tag{15}$$

where $\mathcal{X} = \left\{ \mathbf{p} | \sum_{k=1}^{K} p_k = 1, p_k > 0, k = 1, \ldots, K \right\}$.

There are several ways to carry out perfect sampling from Eq. (16). The first method is based on CFTP [48]. An alternative perfect sampling method for simple mixture models is introduced by [49]. The third approach is to use adaptive rejection sampling [3, 18], since the posterior is log-concave. We may also use the ratio-of-uniform method. However, none of these methods are more efficient than the geometric–arithmetic mean method in [19].

A.1 Geometric-arithmetic mean method

Suppose that $\mathbf{p}^*$, the MLE of $\mathbf{p}$ is unique and for simplicity, assume the prior $\pi_0(\mathbf{p})$ is uniform. Define $a_{nk} = L_{nk} / \sum p_k^* L_{nk}$. The posterior density of $\mathbf{p}$ is then given by

$$f(\mathbf{p}|y) \propto h(\mathbf{p}|y) = \prod_{n=1}^{N} \left(\sum_{k=1}^{K} p_k a_{nk} \right) I_{\mathcal{X}}(\mathbf{p}), \tag{16}$$

where $\mathcal{X}$ is defined in (16).

Let I_n be a random element of $\arg\max_k L_{nk}$. Define $A_j = \{n : I_n = j\}$ and let $\mathbf{n} = (n_1, \ldots, n_k)$ where n_j is the number of elements in A_j.

Define $\mathbf{M} = \{m_{jk}\}$ with $m_{jk} = \left(\sum_{n \in A_j} a_{nk} \right) / n_j$. If $n_j = 0$, then set $m_{jj} = 1$ and $m_{jk} = 0$ for $j \neq k$. We now make two assumptions, which we will return to later on:

A: $\mathbf{M}$ is invertible.

B: The elements of $\mathbf{v} = \left(\mathbf{M}^T\right)^{-1} \mathbf{1}$ are all positive.

Under these assumptions, we will show that the following rejection sampler generates simulated values from the posterior distribution of $\mathbf{p}$. First we define $\mathbf{V}$ to be the diagonal matrix with diagonal elements $\mathbf{v}^T = (v_1 \ldots, v_K)$.

Algorithm 6.1 (GAM sampler)

Sample $\mathbf{q}$ from the Dirichlet distribution with parameter $\mathbf{n} + \mathbf{1}$.	01
Sample U from Unif[0, 1].	02
Calculate $\mathbf{p}$ with $\mathbf{p} = \mathbf{M}^{-1} V^{-1} \mathbf{q}$.	03
If $U \leq h(\mathbf{p}|y) / \prod_{j=1}^{K} \left(q_j / v_j \right)^{n_j}$,	04
Accept $\mathbf{p}$ and stop;	05
else	06
reject $\mathbf{p}$ and go to 01.	07

Proposition 6.1 Under assumptions **A** and **B**, Algorithm 6.1 samples $\mathbf{p}$ with probability density (17).

Proof: Since the geometric average is no larger than the arithmetic average, for $\mathbf{p} \in \mathcal{X}$, we have

$$h(\mathbf{p}|y) = \prod_{n=1}^{N}\left(\sum_{k=1}^{K} p_k a_{nk}\right) = \prod_{j=1}^{K}\prod_{n\in A_j}\left(\sum_{k=1}^{K} p_k a_{nk}\right) \tag{17}$$

$$\leq \prod_{j=1}^{K}\left(\frac{\sum_{n\in A_j}\left(\sum_{k=1}^{K} p_k a_{nk}\right)}{n_j}\right)^{n_j}, \tag{18}$$

where in the case $n_j = 0$, the product term is taken as 1. So that, for $\mathbf{p} \in \mathcal{X}$, with m_{jk} as previously defined, we have

$$h(\mathbf{p}|y) \leq \prod_{j=1}^{K}\left(\sum_{k=1}^{K} p_k m_{jk}\right)^{n_j} \tag{19}$$

$$= \left[\prod_{j=1}^{K} v_j^{-n_j}\right]\prod_{j=1}^{K}\left(v_j\sum_{k=1}^{K} p_k m_{jk}\right)^{n_j} \tag{20}$$

$$= \prod_{j=1}^{K}\left(q_j/v_j\right)^{n_j}, \tag{21}$$

where $q_j = v_j\sum_{k=1}^{K} p_k m_{jk}, j = 1, \ldots, K$ or equivalently $\mathbf{q} = \mathbf{VMp}$.
Since $v_j > 0$ and $\sum_{k=1}^{K} p_k m_{jk} > 0$, it follows that $q_j > 0$ for $j = 1, \ldots, K$. Furthermore

$$\sum_{j=1}^{K} q_j = \sum_{j=1}^{K} v_j \sum_{k=1}^{K} p_k m_{jk} = \mathbf{p}^T\mathbf{M}^T\mathbf{v} = \mathbf{p}^T\mathbf{1} = 1,$$

since $\mathbf{M}^T\mathbf{v} = \mathbf{1}$, from the definition of $\mathbf{v}$. It follows that $\mathbf{p} \in \mathcal{X}$ implies $\mathbf{q} \in \mathcal{X}$, so that

$$h(\mathbf{p}|y)I_{\mathcal{X}}(\mathbf{p}) \leq \prod_{j=1}^{K}\left(q_j/v_j\right)^{n_j} I_{\mathcal{X}}(\mathbf{q}). \tag{22}$$

Note that the right-hand side of Eq. (22) is proportional to a Dirichlet distribution with parameters $(n_1 + 1, \ldots, n_K + 1)$.

Rejection sampling then proceeds as usual:

- A sample $\mathbf{q}$ is drawn from Dirichlet$(\mathbf{n} + \mathbf{1})$.
- The value $\mathbf{p} = \mathbf{M}^{-1}\mathbf{V}^{-1}\mathbf{q}$ is calculated.
- It is accepted with probability $h(\mathbf{p}|y)I_{\mathcal{X}}(\mathbf{p})/\prod_{j=1}^{K}\left(q_j/v_j\right)^{n_j}$.

We now return to assumptions **A** and **B**. Suppose that **M** is invertible but the elements of $\mathbf{v} = \left(\mathbf{M}^T\right)^{-1}\mathbf{1}$ are not all positive. In this case, let

$\alpha_k = \frac{1}{N}\sum_{n=1}^{N} a_{nk}, \quad \alpha = \max_k \{\alpha_k\}, \quad \mathbf{v} = (\alpha N)^{-1}\mathbf{n}$ and $\tilde{\mathbf{M}} = \alpha \mathbf{M} \Delta,$

where Δ is a diagonal matrix with diagonal elements $(1/\alpha_1, \dots, 1/\alpha_K)$. Note that $\mathbf{v} = \left(\tilde{\mathbf{M}}^T\right)^{-1}\mathbf{1} > 0$. Algorithm 6.1 and its proof can then be modified by replacing $\mathbf{M}$ by $\tilde{\mathbf{M}}$.

Suppose now that assumption **A** does not hold, i.e. $\mathbf{M}$ is not invertible. This can be remedied by adding positive quantities to the diagonal elements of $\mathbf{M}$. This also provides an alternative way of ensuring that the elements of $\mathbf{v}$ are positive.

A.2 Dirichlet priors and *pseudo* data

Suppose that the prior $\pi_0(\mathbf{p})$ is Dirichlet$(\alpha_1 + 1, \dots, \alpha_K + 1)$, where $\alpha_i : i = 1, \dots, K$ are positive, integers and let $A = \sum_{j=1}^{K}\alpha_j$. The prior can be synthesized by introducing *pseudo* data, $\tilde{a}_{mk}, m = 1, \dots, A; k = 1, \dots, K$, defined as follows:

$$\tilde{a}_{mk} = \begin{cases} 1 & \text{if} \sum_{j=1}^{k-1}\alpha_j + 1 \le m \le \sum_{j=1}^{k}\alpha_j \\ 0 & \text{otherwise,} \end{cases} \tag{23}$$

since

$$\pi_0(\mathbf{p}) \propto \prod_{k=1}^{K} p_k^{\alpha_k} = \prod_{m=1}^{A}\left(\sum_{k=1}^{K}\tilde{a}_{mk}p_k\right). \tag{24}$$

With the Dirichlet prior, the posterior distribution given by Eq. (16) can be written as

$$f(\mathbf{p}|y) \propto \prod_{l=1}^{N+A}\left(\sum_{k=1}^{K} p_k \overline{a}_{lk}\right) I_{\mathcal{X}}(\mathbf{p}), \tag{25}$$

where $\{\overline{a}_{lk}, l = 1, \dots, N + A\}$ contains the real data $\{a_{nk}, n = 1, \dots, N\}$ and the pseudo data $\{\tilde{a}_{mk}, m = 1, \dots, A\}$.

The posterior distribution (26) has the same form as Eq. (17). Therefore GAM can be used to sample realizations from the posterior distribution (26).

A.3 Simulation results and discussion

We compare the running time of mixture models with sample sizes (N) and different number of components (K) in **Table 1**. The components have specified normal distributions with means $\mu = (\mu_1, \dots, \mu_K)$ and variances $\sigma^2 = (\sigma_1^2, \dots, \sigma_K^2)$. The prior on $\mathbf{p}$ is uniform. We sample 10,000 realizations from the posterior of the models.

When $K = 3, 4$, we simulate N observations from a three-component normal mixture with $\mu = (0, 0, 2)$, $\sigma^2 = (1, 4, 1)$ and mixture weight $\mathbf{p}_0 = (1/2, 1/3, 1/6)$. We then either sample from the posterior distribution of $\mathbf{p}$ using the same distributional components in the case $K = 3$ or sample from the posterior distribution of $\mathbf{p}$ with an additional component having mean $\mu_4 = 4$ and variance $\sigma_4^2 = 4$, in the case $K = 4$.

When $K = 5$, observations are simulated from the normal mixture distribution with components having means $\mu = (-2, 0, 4, 2, 3)$, variances $\sigma^2 = (1, 1, 4, 1, 4)$

and $\mathbf{p}_0 = (0.35, 0.3, 0.1, 0.2, 0.05)$. Samples from the posterior distribution of $\mathbf{p}$ are drawn assuming the same components. Similar calculations are carried out for $K = 6$, where $\mu = (0, 3, 2, -2, -4, 5)$, variances $\sigma^2 = (1, 1, 1, 1, 1, 4)$ and $\mathbf{p}_0 = (0.05, 0.3, 0.3, 0.1, 0.08, 0.17)$, again assuming that the component distributions are known.

From **Table 1**, we can see that the GAM algorithm, while using very little memory, is highly efficient in running time. The last row of the table is the estimated acceptance probability of the GAM algorithm. The algorithm is very efficient when the component densities are known. We can see this not only by simulation but also from theoretical considerations, as follows.

A.3.1 Explanation of efficiency

When $\mathbf{v} = \left(\mathbf{M}^T\right)^{-1}\mathbf{1} > 0$, we are able to use $\mathbf{M}$ directly without modification to construct the hat function, thereby speeding up the calculations. In the simulations of the previous section, this was always found to be the case. Now we explain why this should be so.

If the maximum likelihood estimator of $\mathbf{p}$ is consistent, then when $N \to \infty$,

$$\frac{1}{n_j}\sum_{n \in A_j}\left|\frac{L_{nk}}{\sum_{k=1}^{K} p_k^* L_{nk}} - \frac{L_{nk}}{\sum_{k=1}^{K} p_k L_{nk}}\right| \xrightarrow{p} 0. \tag{26}$$

Assuming sufficient regularity, we also have

$$m_{jk} = \frac{\sum_{n \in A_j} a_{nk}}{n_j} \tag{27}$$

$$= \frac{1}{n_j}\sum_{n \in A_j}\frac{L_{nk}}{\sum_{k=1}^{K} p_k^* L_{nk}} \tag{28}$$

$$\xrightarrow{p} E\left(\frac{f_k(Y)}{f(Y)}|\mathcal{L}_j\right) \tag{29}$$

$$= \frac{\int_{\mathcal{L}_j} f_k(y)dy}{\gamma_j}, \quad \text{as } N \to \infty, \tag{30}$$

K	3	3	4	4	6	6
N	400	1000	400	1000	400	1000
Fearnhead's	242 s	3610 s	*	*	*	*
Leydold's	≤1 s	3.6 s	*	*	*	*
RoU	16:11 s	28:16 s	31:18 s	68:33 s	88:60 s	152:76 s
GAM	4 s	9 s	11 s	16 s	6 s	11 s
GAM AP	0.7472	0.7509	0.2433	0.3088	0.5325	0.5505

*Fearnhead's algorithm, Leydold's algorithm and ratio-of-uniform. GAM method. GAM acceptance probability. The * indicates that Fearnhead's method, and Leydold's method will not run on a standard desktop when K = 4 and K = 5.*

Table 1.
Running times (in s).

where Y has density $f(y) = \sum_{k=1}^{K} p_k f_k(y)$, $\mathcal{L}_j = \left\{ y | f_j(y) \geq f_k(y), k = 1, \dots, K \right\}$ and $\gamma_j = \int_{\mathcal{L}_j} f(y)dy$.

Therefore

$$\mathbf{M}^T \xrightarrow{p} \mathbf{W}^T, \tag{31}$$

where

$$\mathbf{W} = \left\{ w_{jk} \right\}, w_{jk} = \frac{\int_{\mathcal{L}_j} f_k(y)dy}{\gamma_j}. \tag{32}$$

Let $\overline{\mathbf{v}} = (\gamma_1, \dots, \gamma_K)$, then

$$\mathbf{W}^T \overline{\mathbf{v}} = \begin{bmatrix} \sum_{j=1}^{K} \int_{\mathcal{L}_j} f_1(y)dy \\ \vdots \\ \sum_{j=1}^{K} \int_{\mathcal{L}_j} f_K(y)dy \end{bmatrix} = \begin{bmatrix} \int f_1(y)dy \\ \vdots \\ \int f_K(y)dy \end{bmatrix} = \mathbf{1}, \tag{33}$$

where the second equal sign is because $\cup_{j=1}^{K} \mathcal{L}_j = (-\infty, \infty)$. So, there exists $\overline{\mathbf{v}} > 0$, satisfying $\mathbf{W}^T \overline{\mathbf{v}} = \mathbf{1}$. Using Eq. (32), we can conclude that when N is large enough, there also exists $\mathbf{v} \approx \overline{\mathbf{v}} > 0$, satisfying $\mathbf{M}^T \mathbf{v} = \mathbf{1}$.

Since $\gamma_j = \int_{\mathcal{L}_j} f(y)dy$ and $n_j = \#\{A_j\}$, we have $n_j/N \xrightarrow{p} \gamma_j$. When each $n_j, j = 1, \dots, K$ is large, if a random sample $\mathbf{q}$ is drawn from a Dirichlet distribution with parameter $\mathbf{n} + \mathbf{1}$, then each q_j is approximately equal to $n_j/N \approx \gamma_j$. Furthermore, $v_j \approx \gamma_j$, so $\mathbf{q}$ satisfies

$$\mathbf{V}^{-1}\mathbf{q} \approx \mathbf{1}, \tag{34}$$

and then,

$$\mathbf{p} = \mathbf{M}^{-1}\mathbf{V}^{-1}\mathbf{q} \approx \mathbf{M}^{-1}\mathbf{1} = \mathbf{p}^*. \tag{35}$$

If $\mathbf{p}$ is approximately equal to the mode $\mathbf{p}^*$, the two sides of the inequality,

$$h(\mathbf{p}|y) = \prod_{n=1}^{N} \left(\sum_{k=1}^{K} p_k a_{nk} \right) \leq \left[\prod_{j=1}^{K} v_j^{-n_j} \right] \prod_{j=1}^{K} q_j^{n_j}, \tag{36}$$

are approximately equal as well. Thus, the closer the sampled realization $\mathbf{p}$ is to $\mathbf{p}^*$, the larger the acceptance probability is. So the algorithm runs very rapidly, since the sampled values of $\mathbf{p}$ are always around the mode $\mathbf{p}^*$.

This algorithm requires calculating the MLE, which can be performed very quickly since the likelihood function is log-concave. In fact an approximate *guess* for $\mathbf{p}^*$ will suffice. The more accurate the guess is, the more efficient the algorithm will be.

The method performs well when the component densities are correctly specified, as explained in the previous section. For these same reasons, we would expect the algorithm to perform poorly under misspecification. Details of robustness to misspecification can be found in [19].

Author details

Hongsheng Dai
University of Essex, Colchester, UK

*Address all correspondence to: hdaia@essex.ac.uk

References

[1] Cowles MK, Carlin BP. Markov chain Monte Carlo convergence diagnostics: A comparative review. Journal of the American Statistical Association. 1996; **91**:883-904

[2] Fill JA. An interruptible algorithm for perfect sampling via Markov chains. The Annals of Applied Probability. 1998;**8**:131-162

[3] Gilks WR, Wild P. Adaptive rejection sampling for Gibbs sampling. Applied Statistics. 1992;**41**:337-348

[4] Propp JG, Wilson DB. Exact sampling with coupled Markov chains and applications to statistical mechanics. Random Structures and Algorithms. 1996;**9**:223-252

[5] Fill JA, Machida M, Murdoch DJ, Rosenthal JS. Extension of Fill's perfect rejection sampling algorithm to general chains. Random Structure and Algorithms. 2000;**17**:290-316

[6] Propp JG, Wilson DB. How to get an exact sample from a generic Markov chain and sample a random spanning tree from a directed graph, both within the cover time. Journal of Algorithms. 1998;**27**:170-217

[7] Aldous DJ. The random walk construction of uniform spanning trees and uniform labelled trees. SIAM Journal on Discrete Mathematics. 1990;**3**:450-465

[8] Wilson DB. Generating random spanning trees more quickly than the cover time. In: Annual ACM Symposium on Theory of Computing Archive, Proceedings of the Twenty-Eighth Annual ACM Symposium on Theory of Computing. 1996. pp. 296-303

[9] Beskos A, Papaspiliopoulos O, Roberts GO, Fearnhead P. Exact and computationally efficient likelihood-based estimation for discretely observed diffusion processess. Journal of Royal Statistical Society, B. 2006;**68**:333-382

[10] Beskos A, Roberts GO. Exact simulation on diffusions. The Annals of Applied Probability. 2005;**15**: 2422-2444

[11] Beskos A, Papaspiliopoulos O, Roberts GO. A factorisation of diffusion measure and finite sample path constructions. Methodology and Computing in Applied Probability. 2008;**10**:85-104

[12] Dai H. A new rejection sampling method without using hat function. Bernoulli. 2017;**23**:2434-2465

[13] Dai H, Pollock M, Roberts G. Monte Carlo fusion. Journal of Applied Probability. 2019;**56**:174-191

[14] Tan A, Doss H, Hobert JP. Honest importance sampling with multiple Markov chains. Journal of Computational and Graphical Statistics. 2015;**24**(3):792-826

[15] Green PJ, Murdoch DJ. Exact sampling for Bayesian inference: Towards general purpose algorithms. Bayesian Statistics. 1998;**6**:301-321

[16] Wakefield JC, Gelfand AE, Smith AFM. Efficient generation of random variates via the ratio-of-uniforms method. Statistics and Computing. 1991;**1**:129-133

[17] Robert C, Casella G. Monte Carlo Statistical Methods. Springer Science & Business Media; 2013

[18] Leydold J. A rejection technique for sampling from log-concave multivariate distributions. Modeling and Computer Simulation. 1998;**8**(3):254-280. Available from: citeseer.nj.nec.com/leydold98rejection.html

[19] Dai H. Perfect simulation methods for Bayesian applications [PhD thesis], University of Oxford. 2007

[20] Wilson DB. How to couple from the past using a read-once source of randomness. Random Structures and Algorithms. 2000;**16**:85-113

[21] Cai H. Exact sampling using auxiliary variables. In: Statistical Computing Section of ASA Proceedings. 1999

[22] Corcoran JN, Tweedie RL. Perfect sampling from independent Metropolis-Hastings chains. Journal of Statistical Planning and Inference. 2000;**104**(2): 297-314

[23] Mira A, Møller J, Roberts GO. Perfect slice samplers. Journal of the Royal Statistical Society B. 2001;**63**:593-606

[24] Murdoch DJ, Green PJ. Exact sampling from a continuous state space. Scandinavian Journal of Statistics. 1998; **25**:483-502

[25] Hörmann W, Leydold J. Improved perfect slice sampling. Technique Report. 2003

[26] Casella G, Mengersen KL, Robert CP, Titterington DM. Perfect samplers for mixtures of distributions. Journal of the Royal Statistical Society B. 2002;**64**:777-790

[27] Phillipe A, Robert CP. Perfect simulation of positive Gaussian distributions. Statistics and Computing. 2003;**13**:179-186

[28] Huber M. Perfect sampling using bounding chains. The Annals of Applied Probability. 2004;**14**:734-753

[29] Møller J. Perfect simulation of conditionally specified models. Journal of the Royal Statistical Society, B. 1999; **61**:251-264

[30] Kendall WS, Moller J. Perfect simulation using dominating processes on ordered spaces, with application to locally stable point processes. Advances in Applied Probability. 2000;**32**(3): 844-865

[31] Fearnhead P. Perfect simulation from population genetic models with selection. Theoretical Population Biology. 2001;**59**(4):263-279

[32] Dai H. Exact Monte Carlo simulation for fork-join networks. Advances in Applied Probability. 2011;**43**(2):484-503

[33] Dai H. Exact simulation for fork-join networks with heterogeneous service. International Journal of Statistics and Probability. 2015;**4**(1):19-32

[34] Connor SB, Kendal WS. Perfect simulation of M/G/c queues. Advances in Applied Probability. 2015;**47**(4): 1039-1063

[35] Djuric PM, Huang Y, Ghirmai T. Perfect sampling: A review and applications to signal processing. IEEE Transactions on Signal Processing. 2002; **50**(2):345-356

[36] Wilson DB. Layered multishift coupling for use in perfect sampling algorithms (with a primer on CFTP). In: Madras N, editor. Monte Carlo Methods. Vol. 26. Fields Institute Communications. American Mathematical Society; 2000. pp. 141-176

[37] Hansen NR. Geometric ergodicity of discrete-time approximations to multivariate diffusions. Bernoulli. 2003; **9**(4):725-743

[38] Dai H. Exact simulation for diffusion bridges: An adaptive approach. Journal of Applied Probability. 2014;**51**(2):346-358

[39] Agarwal A, Duchi JC. Distributed delayed stochastic optimization. In: 51st

IEEE Conference on Decision and Control; Maui Hawaii, USA. 2012

[40] Li C, Srivastava S, Dunson DB. Simple, scalable and accurate posterior interval estimation. Biometrika. 2017; **104**(3):665-680

[41] Minsker S, Srivastava S, Lin L, Dunson DB. Scalable and robust Bayesian inference via the median posterior. In: Proceedings of the 31st International Conference on Machine Learning (ICML-14). 2014. pp. 1656-1664

[42] Neiswanger W, Wang C, Xing E. Asymptotically exact, embarrassingly parallel MCMC. In: Proceedings of the Thirtieth Conference on Uncertainty in Artificial Intelligence (2014). 2014. pp. 623-632

[43] Scott SL, Blocker AW, Bonassi FV, Chipman HA, George EI, McCulloch RE. Bayes and big data: The consensus Monte Carlo algorithm. International Journal of Management Science and Engineering Management. 2016;**11**(2):78-88

[44] Srivastava S, Cevher V, Tan-Dinh Q, Dunson DB. WASP: Scalable Bayes via barycenters of subset posteriors. In: Proceedings of the Eighteenth International Conference on Artificial Intelligence and Statistics (AISTATS 2016). 2016. pp. 912-920

[45] Stamatakis A, Aberer AJ. Novel parallelization schemes for large-scale likelihood-based phylogenetic inference. In: 2013 IEEE 27th International Symposium on Parallel and Distributed Processing. 2013. DOI: 10.1109/IPDPS.2013.70

[46] Wang X, Dunson. Parallelizing MCMC via Weierstrass Sampler. arXiv preprint arXiv:1312.4605. 2013

[47] Uney M, Clark DE, Julier SJ. Distributed fusion of PHD filters via exponential mixture densities. IEEE Journal of Selected Topics in Signal Processing. 2013;7(3):521-531

[48] Hobert J, Robert C, Titterington D. On perfect simulation for some mixtures of distributions. Statistics and Computing. 1999;**9**:287-298

[49] Fearnhead P. Direct simulation for discrete mixture distributions. Statistics and Computing. 2005;**15**(2): 125-133

Section 3

Bayesian Inference for Complicated Data

Chapter 4

Bayesian Analysis for Random Effects Models

Junshan Shen and Catherine C. Liu

Abstract

Random effects models have been widely used to analyze correlated data sets, and Bayesian techniques have emerged as a powerful tool to fit the models. However, there has been scarce literature that systematically reviews and summarizes the recent advances of Bayesian analyses of random effects models. This chapter reviews the use of the Dirichlet process mixture (DPM) prior to approximate the distribution of random errors within the general semiparametric random effects models with parametric random effects for longitudinal data setting and failure time setting separately. In a survival setting with clusters, we propose a new class of nonparametric random effects models which is motivated from the accelerated failure models. We employ a beta process prior to tact clustering and estimation simultaneously. We analyze a new data set integrated from Alzheimer's disease (AD) study to illustrate the presented model and methods.

Keywords: beta process, Dirichlet process mixture, clustered data, longitudinal data, random effects, survival outcome, nonparametric transformation model

1. Introduction

Random effects models have been widely used as a powerful tool for analyzing correlated data [1, 2]. The model features a finite number of random terms acting as latent variables to model unobserved factors; see [3] for a comprehensive review. Some authors have further proposed semiparametric mixed effect models by allowing for infinite dimensional random effects [4, 5]. Most of the aforementioned works draw inferences using frequentist approaches, while Bayesian approaches have been largely ignored because of the lack of computational feasibility and expediency. With the advent of the "supercomputer" era, Bayesian analyses have recently sparked much interest in the setting of random effects models for clustered data or longitudinal settings. However, there is scarce literature that has systematically reviewed the Bayesian works in the area.

By extending the traditional random effects models, recent research focus has shifted to study heterogeneous random effects or nonparametric distributions of random effects, which arise because of skewness of data, missing covariates, or unmeasurable subject-specific covariates [6]. The extended random effects models, termed semiparametric random effects models, improve statistical performance with added interpretability. Bayesian techniques, which provide a convenient means to model non-Gaussian distributions, have recently been proposed for semiparametric random effects model in a variety of settings ([7, 8], among others).

The discreteness of the Dirichlet process makes it impossible as a prior for estimating a density. However, as a remedy by convolving with a kernel, Dirichlet process mixture plays an important role [9].

For censored outcome data, transformation models, which transform the time-to-event responses using a monotone function and link them to the covariates of interest, have surged as a strong competitor of the Cox model [10]. Moreover, the transformation model framework is fairly general. The Cox model and the proportional odd model [11] can be viewed as nonparametric transformation linear models with some specific error terms; see [12–14]. For correlated data, the transformation model naturally extends the semiparametric random effects model by directly incorporating random effects to the transformation functions, treating them as realizations of an underlying random function. Bayesian analyses have found much use in this new area. For example, the beta process has been found to be a reasonable candidate for the prior of the monotone transformation function [15–17].

This chapter focuses on the Bayesian analysis of the transformed linear model with censored data and in a clustered setting. In many biomedical studies, the observations are naturally clustered. For example, patients in observational studies can be grouped in analysis according to a variety of factors, such as age, race, gender, and hospital, in order to reduce the confounding effects. Following Mallick and Walker [18], we explore using a mixture of beta distributions and the beta process as the candidates for the prior distribution of the random transformation function [17, 19, 20].

The rest of this chapter is structured as follows. Section 2 reviews the use of the Bayesian approach to infer parametric random effects models. In the setting of survival analysis, Section 3 proposes a beta process prior to fit random effects model with nonparametric transformation functions, and Section 4 applies the method to study the progression of Alzheimer's disease (AD). Section 5 concludes the chapter with future research directions.

2. Dirichlet process mixture prior

In parametric random effects models, we considered the situation that the distribution form of the random error term is unknown. Dirichlet process mixture (DPM) is used as the prior for the baseline distribution in that error terms used to be continuous random variables in most situations.

2.1 Linear mixed effects model

With a longitudinal data set $\{Y_i, x_i, z_i\}$, we posit a mixed effects model with an AR(1) serial correlation structure:

$$\begin{aligned} \mathbf{y}_i &= \mathbf{x}_i\boldsymbol{\beta} + \mathbf{z}_i\mathbf{b}_i + \mathbf{w}_i, i = 1, \dots, m; \\ \mathbf{w}_i &= (w_{i1}, \dots, w_{in_i})^T; w_{ij} = \rho w_{i,j-1} + \epsilon_{ij}, j = 2, \dots, n_i, \end{aligned} \tag{1}$$

where $\mathbf{y}_i = \left(y_{i1}, \dots, y_{in_i}\right)^T$ with y_{ij} being the jth response of the ith subject for $i = 1, \dots, m$, β is a $p \times 1$ vector of fixed effect parameters, $\mathbf{b}_i$ a $q \times 1$ Gaussian random vector representing the subject-specific random effects, $\mathbf{x}_i$ and $\mathbf{z}_i$ are $n_i \times p$ and $n_i \times q$ design matrices linking β and $\mathbf{b}_i$ to $\mathbf{y}_i$, respectively, $\mathbf{w}_i = (w_{i1}, \dots, w_{in_i})^T$ is an $n_i \times 1$ vector of model errors, ρ is the autoregressive coefficient, and ϵ_{ij},s are i.i.d. noises. When $\{\epsilon_{ij}\}$ is non-normal, we assume a mixture model:

$$f_G(\epsilon|\sigma^2) = \int \varphi(\epsilon|u, \sigma^2) dG(u), \tag{2}$$

where $\varphi(\cdot|u, \sigma^2)$ is the probability density function for a normal random variable with mean u and variance σ^2 and G is an unspecified probability distribution of u satisfying $\int u dG(u) = 0$, which ensures that ϵ comes from a mean-zero mixture distribution.

Replacing the Dirichlet process by an equivalent Pólya urn representation, [8] employed an empirical likelihood approach with the moment constraints and developed a posterior adjusted Gibbs sampler for more precise estimation. The algorithm is computationally feasible.

2.2 Accelerated failure time model

We shift gears to study survival outcomes with a cluster structure. Denote the data set by $(T_{ij}, X_{ij}), i = 1, \cdots, K, j = 1, \cdots, n_i$, where T_{ij} is the failure time of the jth subject in the ith cluster and X_{ij} is a vector of associated covariates. To accommodate such data, we utilize a general accelerated failure time model:

$$\log T_{ij} = \mathbf{X}_{ij}^T \boldsymbol{\beta} + \varepsilon_{ij}, \qquad i = 1, \cdots, K \ \text{ and } \ j = 1, \cdots, n_i, \tag{3}$$

where β is a vector of p-dim regression coefficients of interest and ε_{ij} are independent random errors following the distribution with density f_i. [7] posed an exponential tilt on the distributions of error terms to incorporate the cluster heterogeneity. That is,

$$\frac{f_i(t)}{f_1(t)} = \exp\left(\theta_{0i} + \boldsymbol{\theta}_i^T q(t)\right), \qquad i = 2, \cdots, K, \tag{4}$$

where $q(t)$ is a q-dimensional prespecified functions containing potential covariate information and $\boldsymbol{\theta}_i$ is the corresponding parameter vector with $\theta_{0i} = \log\left[\left(\int \exp(\boldsymbol{\theta}_i^T q(t)) f_1(t) dt\right)^{-1}\right]$. Thus, $\boldsymbol{\theta}_i$ represents the parametric random effects in the model. Li et al. [7] place the DPM prior on the baseline density f_1 to develop a set of procedures which improves estimation efficiency through information pooling.

3. Beta process prior

We now present a nonparametric random effects model for the clustered survival data with nonparametric monotone link functions. We employ a beta process as the prior for the baseline function.

Let T_{ij} denote the failure time of the jth subject in the ith cluster, X_{ij} be the covariate vector for the subject, and C_{ij} be the potential censoring time to the jth subject in the ith cluster. Assume that C_{ij} is independent of the failure time T_{ij}. Let $Z_{ij} = \min(T_{ij}, C_{ij})$ and let $\delta_{ij} = I(T_{ij} < C_{ij})$ be the censoring indicator. Then the observed data can be described as

$$(Z_{ij}, \delta_{ij}, X_{ij}), i = 1, \cdots, n; \quad j = 1, \cdots, n_i. \tag{5}$$

Within each cluster, T_{ij} is linked to X_{ij} via the following transformation model:

$$\ln H_i(T_{ij}) = X_{ij}^T \boldsymbol{\beta} + \ln\varepsilon_{ij}, i = 1, 2\cdots, n, \quad (6)$$

where ε_{ij} are i.i.d. variables with a known density function $f_\varepsilon(\cdot)$ and $H_i(t)$ are unknown cluster-specific monotone functions, which are i.i.d. realizations of a random function and can be viewed as a nonparametric version of random effects for independent clusters. In a parametric setting, if we set $H_i(t) = t\exp(-b_i)$ with b_i being a cluster-specific random effect, Eq. (6) reduces to a classical random effects model, which has been discussed in Section 2.2. The challenge, however, lies in how to draw inferences in such a nonparametric setting.

To proceed, let the coefficient vector $\boldsymbol{\beta}$ be a p-dim unknown vector of interest. We further assume H_i's are differentiable with derivative $h_i(t) = H_i'(t)$, and then the likelihood based on the observed data is

$$L(\boldsymbol{\beta}, H_1, \cdots, H_n | data) = \prod_{i=1}^{n} \prod_{j=1}^{n_i} p(T_{ij}, X_{ij}, \delta_{ij} | H_i, \beta), \quad (7)$$

where

$$p(t, x, \delta | H, \beta) = \left(f_\varepsilon\left(H(t) e^{-x^T\beta}\right) h(t) e^{-x^T\beta}\right)^\delta S_\varepsilon\left(H(t) e^{-x^T\beta}\right)^{1-\delta}.$$

Here S_ε is the survival function of *varepsilon* defined by $S_\varepsilon(s) = P(\varepsilon \geq s)$.

We develop a Bayesian inference procedure based on model (6). We assume that the regression coefficient β follows a normal prior:

$$\beta \sim N_p\left(0, \sigma_\beta^2 I_p\right), \quad (8)$$

where I_p is the $p \times p$ dimensional identity matrix. Since H_i is assumed differentiable, we model it with a kernel convolution:

$$H_i = \int \Phi_\sigma(\cdot - s) dB_i(s),$$

where B is an increasing function and Φ_σ is the zero-mean normal distribution with variance σ^2. Hence, the derivative of H_i is

$$h_i = \int \phi_\sigma(\cdot - s) dB_i(s)$$

with $\phi_\sigma(t) = \frac{1}{\sigma}\phi\left(\frac{t}{\sigma}\right)$. This actually mimics the idea of DPM to smooth beta process by convolution.

We are in a position to select an appropriate stochastic process used as the prior of B_i. Beta process, as studied by [16, 17], is an ideal candidate for the prior of a monotone function. Specifically, beta process $\mathrm{BP}(\gamma, B_0)$ with concentration parameter γ and a base measure B_0 is an increasing Lévy process with independent increments of the form

$$dB(t) \sim \mathrm{Beta}(\gamma dB_0(t), \gamma(1 - dB_0(t))).$$

Teh et al. [20] showed that a sample from $\mathrm{BP}(\gamma, B_0)$ could be represented as

$$B_i(y) = \sum_{l=1}^{\infty} p_{il} I(\theta_{il} \le y), \tag{9}$$

where $p_{il} = \prod_{j=1}^{l} \nu_{il}$ and (θ_{il}, ν_{il}) follows

$$\theta_{il} \sim B_0(\theta), \nu_{il} \sim \mathrm{Beta}(\gamma, 1) \quad l = 1, 2, \cdots.$$

In practice, we need to approximate samples of $\mathrm{BP}(\gamma, B_0)$ with a finite dimensional form. Since beta process $\mathrm{BP}(\gamma B_0)$ can be represented by a stick-breaking process defined in Eq. (9), a natural approximation is obtained by retaining its first L components. That is,

$$B_i^* = \sum_{l=1}^{L} p_{il} \delta_{\theta_{il}},$$

with $p_{il} = \prod_{j=1}^{l} \nu_{il}, l = 1, \cdots, L$. Denote $\boldsymbol{\xi}_i = (\nu_{i1}, \cdots, \nu_{iL}, \theta_{i1}, \cdots, \theta_{iL})^T$ and define

$$H_\sigma^*(z, \boldsymbol{\xi}_i) = \sum_{l=1}^{L} p_{il} \Phi_\sigma(z - \theta_{il}), h_\sigma^*(z, \boldsymbol{\xi}_i) = \sum_{l=1}^{L} p_{il} \phi_\sigma(z - \theta_{il}).$$

The approximated posterior based on the truncated DP is

$$\pi(\beta) \prod_{i=1}^{n} \left[\pi^{\xi}(\boldsymbol{\xi}_i) \prod_{j=1}^{n_i} f(Z_{ij}, X_{ij}, \delta_{ij} | \beta, \boldsymbol{\xi}_i) \right], \tag{10}$$

where

$$f(z, x, \delta | \beta, \boldsymbol{\xi}) = \left(p_\varepsilon\left(H_\sigma^*(z, \boldsymbol{\xi}) \exp\left(-x^T \boldsymbol{\beta}\right)\right) h_\sigma^*(z, \boldsymbol{\xi}) \exp\left(-x^T \boldsymbol{\beta}\right)\right)^\delta$$
$$\times \left(P_\varepsilon\left(H_\sigma^*(z, \boldsymbol{\xi}) \exp\left(-x^T \boldsymbol{\beta}\right)\right)\right)^{1-\delta}.$$

The samples for $\boldsymbol{\beta}$ and $(\boldsymbol{\xi}_1, \ldots, \boldsymbol{\xi}_n)$ based on the posterior can be obtained with Markov chain Monte Carlo (MCMC) [21]. In our simulation, we use the R-package MCMC (https://cran.r-project.org/web/packages/mcmc/index.html) to draw samples for $\boldsymbol{\xi}_1, \ldots, \boldsymbol{\xi}_n$ and β and use the Metropolis algorithm with a normal working distribution.

4. An application to Alzheimer's disease neuroimaging initiative

Alzheimer's Disease Neuroimaging Initiative (ADNI) is a multisite cooperative study for the purpose of improving the prevention and treatment of Alzheimer's disease. The subjects in the study fall into three groups, cognitively normal (CN) individuals, mild cognitive impairment (MCI) patients, and early AD patients. ADNI provides a rich array of patients' information, including functional magnetic resonance imaging (fMRI), positron emission tomography (PET), longitudinal functional cognitive tests scores, blood samples, genetics data, and censored failure time outcomes. Details of the study can be found at http://adni.loni.usc.edu.

We focus on the MCI group. MCI is recognized as a transitional stage between normal cognition and Alzheimer's disease. The failure time is defined to be the time that a MCI patient is diagnosed with AD, which will be censored if a MCI patient remains at the MCI stage at the end of the follow-up time. Wide heterogeneities are exhibited among the failure times, which may be due to demographics and a variety of functional clinical biomarkers, such as the brain areas of the hippocampus, ventricles, and entorhinal cortex. The goal of the analysis is to study the impact of risk factors on progression to AD.

Using the same data as analyzed by [14], we demonstrate our methodology by modeling the failure time (the observed time of AD diagnosis from MCI stage in year) of 281 MCI patients on gender (0 = female, 1 = male), years of education, the number of apolipoprotein E alleles (0, 1, or 2), and the baseline hippocampal volume.

As age is a strong confounder but the functional form of its impact has not reached consensus, we elect to model its impact nonparametrically. Specifically, we use age to form two strata (below and above the median age) and use model (6) to estimate the stratum-specific transformation functions and the effects of other covariates. For comparisons, we also fit model (6) with age as a continuous variable and with a common transformation function. That is, we do not assume the data are clustered. For both models, the regression errors ε's are assumed to follow an exponential distribution with mean 10. In our calculation, we approximate the BPs by a finite truncation with $L = 20$. We assume the precision parameter $\alpha = 1$ and scale parameter $\sigma^2 \sim 1/\sigma^2$.

Figure 1 illustrates the estimated transformation function H of the failure time without clustering. The posterior means (PM) and standard errors (SE) of the regression coefficients in the model are reported in **Table 1**. We run the MCMC for

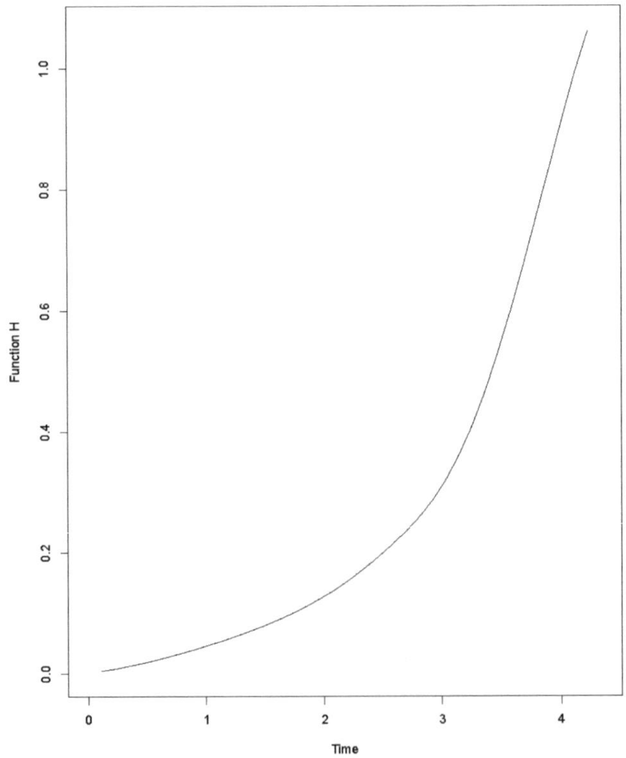

Figure 1.
Smoothed transformation function without clustering.

20,000 iterations with the first 4000 draws discarded as burn-in samples and use Geweke's statistic to ensure the convergence of the chains.

Figure 2 illustrates the estimated transformation functions with age-stratified data, and **Table 2** summarizes the posterior means and standard errors of the other regression coefficients.

The left curve is relatively flat, while the right curve has a sharper slope. This is consistent with the recognition that AD is an aging disease: elder people above a certain age threshold tend to progress faster from MCI to AD.

Both **Tables 1** and **2** show that none of the biomarkers are significant, whereas they are statistically significant in the analysis of [14]. One possible conjecture is that our nonparametric transformation functions may have well captured the effects of unobserved confounders, which may leave little to be explained by the observed covariates. More thorough investigation is warranted.

	RID	AGE	PTGENDER	PTEDUCAT	APOE4	Hipp.
PM	−0.9635	0.0069	−0.1453	−0.0231	−0.1817	0.2710
SE	1.3288	0.0841	1.2331	0.1835	0.8616	0.5333

Table 1.
Posterior estimates of regression coefficients with standard errors.

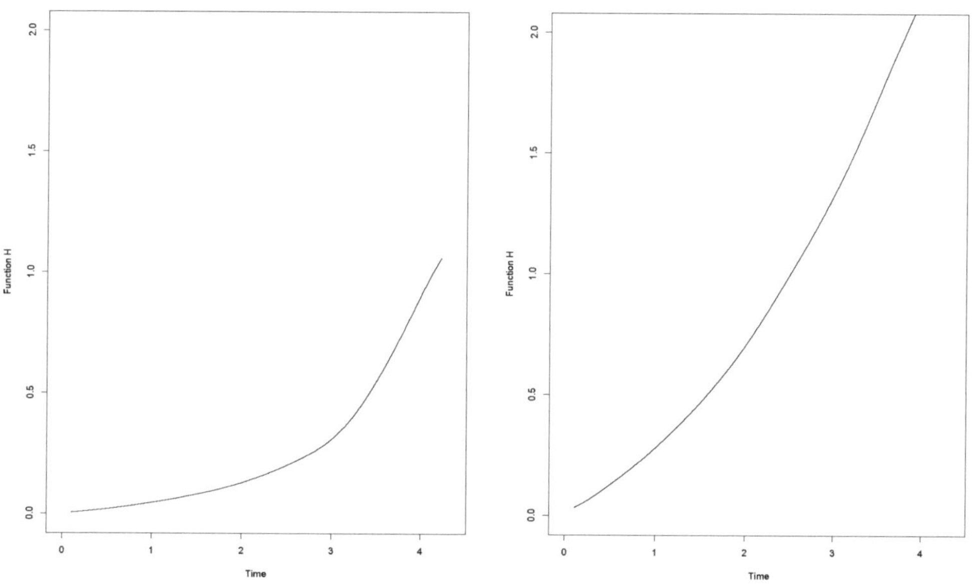

Figure 2.
Smoothed transformation functions with two age-strata: The left curve is the smoothed transformation function for group aged below the average age; the right curve is the smoothed transformation function for the group aged over the average age.

	RID	PTGENDER	PTEDUCAT	APOE4	Hipp.
PM	−0.6399	−0.0706	−0.0072	−0.1349	0.1919
SE	0.9273	0.8491	0.1267	0.6098	0.3716

Table 2.
Posterior estimators of regression coefficients with standard errors.

5. Future directions

Following [12], we can extend the transformation model (6) by allowing the error function f_ε to be unspecified. In this case, we need to specify the regression coefficient β to obey some constraints such as $\beta_1 = 1$ or $\|\boldsymbol{\beta}\| = 1$ for identifiability. We will propose to model the error function using a Dirichlet processes mixture model:

$$f_\varepsilon(t) = \int \varphi(t|\mu, \sigma^2) dG(\mu, \sigma^2), \quad G \sim \mathrm{DP}\left(\alpha, G_0 = \mathrm{N}\left(\mu|\mu_0, \sigma_0^2\right) \times \mathrm{IG}\left(\alpha_1, \alpha_2\right)\right),$$

where $\varphi(t|\mu, \sigma^2)$ is a normal kernel with mean μ and variance σ^2 and G are samples from a Dirichlet process $\mathrm{DP}\left(\alpha_1, G_0 = \mathrm{N}\left(\mu|\mu_0, \sigma_0^2\right) \times \mathrm{IG}\left(a, b\right)\right)$, where α_1 is the mass parameter and $\mathrm{IG}(\cdot|a, b)$ is the inverse gamma distribution with shape parameter a and scale parameter b.

In a slightly different context, we may also consider clustering observations by developing a new nested beta-Dirichlet process prior with companion MCMC algorithms. As there are limited works on functional random effects models that accommodate clustering structures observed, for example, from neural studies, we may propose a nested Dirichlet process [19] as the prior of Dirichlet process to cluster cumulative distribution functions successfully. We envision that such a nested Bayesian procedure will provide substantial computational expedience for practitioners and can certainly be applied to studies that cover beyond the neurodegenerative and aging diseases.

Acknowledgements

Shen's research is partially supported by Beijing Natural Science Foundation 1192006 and National Natural Science Foundation of China; Liu's research is partially supported by General Research Fund, Research Grants Council, Hong Kong, 15327216, and the Hong Kong Polytechnic University grant YBTR. Data collection and sharing for this project were funded by the Alzheimer's Disease Neuroimaging Initiative (ADNI) (National Institutes of Health Grant U01 AG024904) and DOD ADNI (Department of Defense award number W81XWH-12-2-0012). ADNI is funded by the National Institute on Aging, the National Institute of Biomedical Imaging and Bioengineering, and through generous contributions from the following: AbbVie, Alzheimer's Association; Alzheimer's Drug Discovery Foundation; Araclon Biotech; BioClinica, Inc.; Biogen; Bristol-Myers Squibb Company; CereSpir, Inc.; Cogstate; Eisai Inc.; Elan Pharmaceuticals, Inc.; Eli Lilly and Company; EuroImmun; F. Hoffmann-La Roche Ltd and its affiliated company Genentech, Inc.; Fujirebio; GE Healthcare; IXICO Ltd.; Janssen Alzheimer Immunotherapy Research Development, LLC.; Johnson & Johnson Pharmaceutical Research Development LLC.; Lumosity; Lundbeck; Merck Co., Inc.; Meso Scale Diagnostics, LLC.; NeuroRx Research; Neurotrack Technologies; Novartis Pharmaceuticals Corporation; Pfizer Inc.; Piramal Imaging; Servier; Takeda Pharmaceutical Company; and Transition Therapeutics. The Canadian Institutes of Health Research is providing funds to support ADNI clinical sites in Canada. Private sector contributions are facilitated by the Foundation for the National Institutes of Health (www.fnih.org). The grantee organization is the Northern California Institute for Research and Education, and the study is coordinated by the Alzheimer's Therapeutic Research Institute at the University of Southern California. ADNI data are disseminated by the Laboratory for Neuro Imaging at the University of Southern California.

Data used in preparation of this article were obtained from the Alzheimers Disease Neuroimaging Initiative (ADNI) database (adni.loni.usc.edu). As such, the investigators within the ADNI contributed to the design and implementation of ADNI and/or provided data but did not participate in analysis or writing of this report. A complete listing of ADNI investigators can be found at: http://adni.loni.usc.edu/wp-content/uploads/how_to_apply/ADNI_Acknowledgement_List.pdf.

Author details

Junshan Shen[1†] and Catherine C. Liu[2*†]

1 Capital University of Economics and Trade, Beijing, China

2 The Hong Kong Polytechnic University, Hong Kong SAR

*Address all correspondence to: macliu@polyu.edu.hk

† These authors contributed equally.

References

[1] Harville DA. Extension of the Gauss–Markov theorem to include the estimation of random effect. The Annals of Statistics. 1976;**4**:384-395

[2] Laird NM, Ware JH. Random-effects models for longitudinal data. Biometrics. 1982;**38**(2):963-974

[3] Li Y. Random effect models. In: Pham H, editor. Springer Handbook of Engineering Statistics. London: Spring-Verlag; 2006. pp. 687-704

[4] Zeger SL, Diggle PJ. Semiparametric models for longitudinal data with application to CD4 cell numbers in HIV seroconverters. Biometrics. 1994;**50**(3): 689-699

[5] Li Y, Lin X, Müller P. Bayesian inference in semiparametric mixed models for longitudinal data. Biometrics. 2010;**66**(1):70-78

[6] Li Y, Mller P, Lin X. Center-adjusted inference for a nonparametric Bayesian random effect distribution. Statistica Sinica. 2011;**21**:1201-1223

[7] Li Z, Xu X, Shen J. Semiparametric Bayesian analysis of accelerated failure time models with cluster structures. Statistics in Medicine. 2017;**36**(25): 3976-3989

[8] Shen J, Yu H, Yang J, Liu C. Semiparametric Bayesian analysis for longitudinal mixed effects models with non-normal AR(1) errors. Statistics and Computing. 2019;**29**(3):571-583

[9] Ghosal S, van der vaart A. Fundamentals of Nonparametric Bayesian Inference. Cambridge: Cambridge University Press; 2017

[10] Cox DR. Regression models and life tables (with discussion). Journal of the Royal Statistical Society: Series B. 1972; **34**(2):187-220

[11] Hanson T, Yang M. Bayesian semiparametric proportional odds models. Biometrics. 2007;**63**(1):88

[12] Horowitz JL. Semiparametric estimation of a regression model with an unknown transformation of the dependent variable. Econometrica. 1996;**64**(1):103-137

[13] Linton O, Sperlich S, Keilegom IV. Estimation of a semiparametric transformation model. Annals of Statistics. 2008;**36**(2):686-718

[14] Li K, Luo S. Functional joint model for longitudinal and time-to-event data: An application to Alzheimer's disease. Statistics in Medicine. 2017;**36**(25): 3560-3572

[15] Müller P, Mitra R. Bayesian nonparametric inference why and how. Bayesian Analysis. 2013;**8**(2):269-302

[16] Kalbfleisch JD. Non-parametric bayesian analysis of survival time data. Journal of the Royal Statistical Society. 1978;**40**(2):214-221

[17] Hjort N. Nonparametric bayes estimators based on beta processes in models for life history data. Annals of Statistics. 1990;**18**(3):1259-1294

[18] Mallick B, Walker S. A Bayesian semiparametric transformation model incorporating frailties. Journal of Statistical Planning and Inference. 2007; **112**(1):159-174

[19] Rodriguez A, Dunson DB, Gelfand AE. The nested Dirichlet process. Journal of the American Statistical Association. 2008;**103**(483): 1131-1154

[20] Teh YW, Görür D, Ghahramani Z. Stick-breaking construction for the Indian buffet process. In: Proceedings of

the Eleventh International Conference on Artificial Intelligence and Statistics; Vol. 2; 2007. pp. 556-563

[21] Geyer CJ. Introduction to Markov chain Monte Carlo. In: Brooks S, Gelman A, Jones G, Meng X-L, editors. Handbook of Markov Chain Monte Carlo. Boca Raton: Chapman and Hall/ CRC; 2011. pp. 3-48

Chapter 5

Bayesian Inference of Gene Regulatory Network

Xi Chen and Jianhua Xuan

Abstract

Gene regulatory networks (GRN) have been studied by computational scientists and biologists over 20 years to gain a fine map of gene functions. With large-scale genomic and epigenetic data generated under diverse cells, tissues, and diseases, the integrative analysis of multi-omics data plays a key role in identifying casual genes in human disease development. Bayesian inference (or integration) has been successfully applied to inferring GRNs. Learning a posterior distribution than making a single-value prediction of model parameter makes Bayesian inference a more robust approach to identify GRN from noisy biomedical observations. Moreover, given multi-omics data as input and a large number of model parameters to estimate, the automatic preference of Bayesian inference for simple models that sufficiently explain data without unnecessary complexity ensures fast convergence to reliable results. In this chapter, we introduced GRN modeling using hierarchical Bayesian network and then used Gibbs sampling to identify network variables. We applied this model to breast cancer data and identified genes relevant to breast cancer recurrence. In the end, we discussed the potential of Bayesian inference as well as Bayesian deep learning for large-scale and complex GRN inference.

Keywords: gene regulatory network, data integration, Bayesian inference, Gibbs sampling, breast cancer

1. Introduction

The era of "big data" has arrived to the field of computational biology [1]. Biological systems are so complex that in many situations, it is not feasible to directly measure the target signals. Actually, most of biological measurements are noisy and dependent to but not exactly about what we aim to find. This is where probability theory comes to our aid: estimate the true signals from noisy measurements in the presence of uncertainty. Bayesian inference has been widely applied in computational biology field. In certain systems for which we have a good understanding, i.e., gene regulation, behind the observed signals, there exist multiple hidden factors controlling how genes behave under a specific condition. As we are lacking observations on those hidden factors, we model them as parameters in a Bayesian framework, with or without informative prior. Then, for each parameter, Bayesian inference learns a "posterior" distribution, through which we make a final estimation with a confidence interval.

Bayesian inference can update the shape of the learned posterior distributions for model parameters whenever new data observations arrive, providing enough flexibility for integrative analysis and model extension [2]. Although using more data types means defining more model parameters, Bayesian inference automatically prefers for simple models that sufficiently explain data without unnecessary complexity. This is a very important property for biological data analysis because a simple model is much easier to validate using lab-controlled experiments.

In this chapter, we introduce how to apply Bayesian inference to inferring gene regulatory networks (GRN). GRN is a hierarchical network with regulatory proteins, target genes, and interactions between them [3], playing a key role in mediating cellular functions and signaling pathways in cells [4]. Accurate inference of GRN using data specific for a disease returns disease-associated regulatory proteins and genes, serving as potential targets for drug treatment [5]. In recent years, noncoding DNA analysis reveals more and more noncoding regions with strong regulatory effects on gene transcription [6], which greatly expands the scope of GRN research.

GRN analysis requires an integration of multiple types of measurements including but not limited to gene expression, chromatin accessibility, transcription factor binding, methylation, and histone modification [7]. The challenge of GRN inference is that there exit hundreds of proteins and tens of thousands of genes. One protein can regulate hundreds of target genes, and their regulatory relationship (an interaction in GRN) may vary across different cell types, tissues, or diseases. Experiments of high-throughput target gene measurements for one protein in one specific condition are costive and noisy [8], let alone for hundreds of proteins under diverse conditions. For many tissues or diseases, we need to integrate multiple relevant data types and computationally infer GRNs specific for those conditions.

Bayesian inference is particularly suitable for GRN inference as it is very flexible for large-scale data integration. Moreover, when we have multiple datasets generated from very similar conditions, estimating variables using distribution learning than a single-value prediction makes the final estimation more robust and easier to compare across multiple datasets. We demonstrated this using two breast cancer datasets generated under very similar conditions, in which we also compared a hierarchical Bayesian model with several competing methods. Moreover, using patient data as model input, although they are noisy, we successfully identified a GRN associated with breast cancer recurrence. Finally, we discussed the potential of Bayesian deep learning for large-scale and complex GRN inference.

2. Gene regulatory networks

Human genome can be simply divided into coding (exomes) and noncoding regions. The process of producing an RNA copy from exomes is called transcription, which can be quantitatively measured using microarray or RNA-seq technics [9, 10], producing gene expression data of ~30,000 genes simultaneously. The transcription process is mediated by regulatory regions located in the noncoding genome, including promoters and enhancers [11]. Promoters are proximal to gene transcription starting sites (TSS), usually within 3 kbps (**Figure 1A**), while enhancers are usually located distantly, i.e., 200 kbps (**Figure 1B**), and can be up to 1 Mbps. In general, each gene could be associated with one promoter and multiple enhancers.

Transcription factors (TFs), a special category of proteins, often coordinate with each other as cis-regulatory modules (CRMs) [12] and co-bind at regulatory regions [13]. For example, in **Figure 1A** or **B**, there are three TFs binding at promoter or enhancer regions and functioning together as one CRM to mediate the transcription process of their target genes. It has been known that the association relationships of

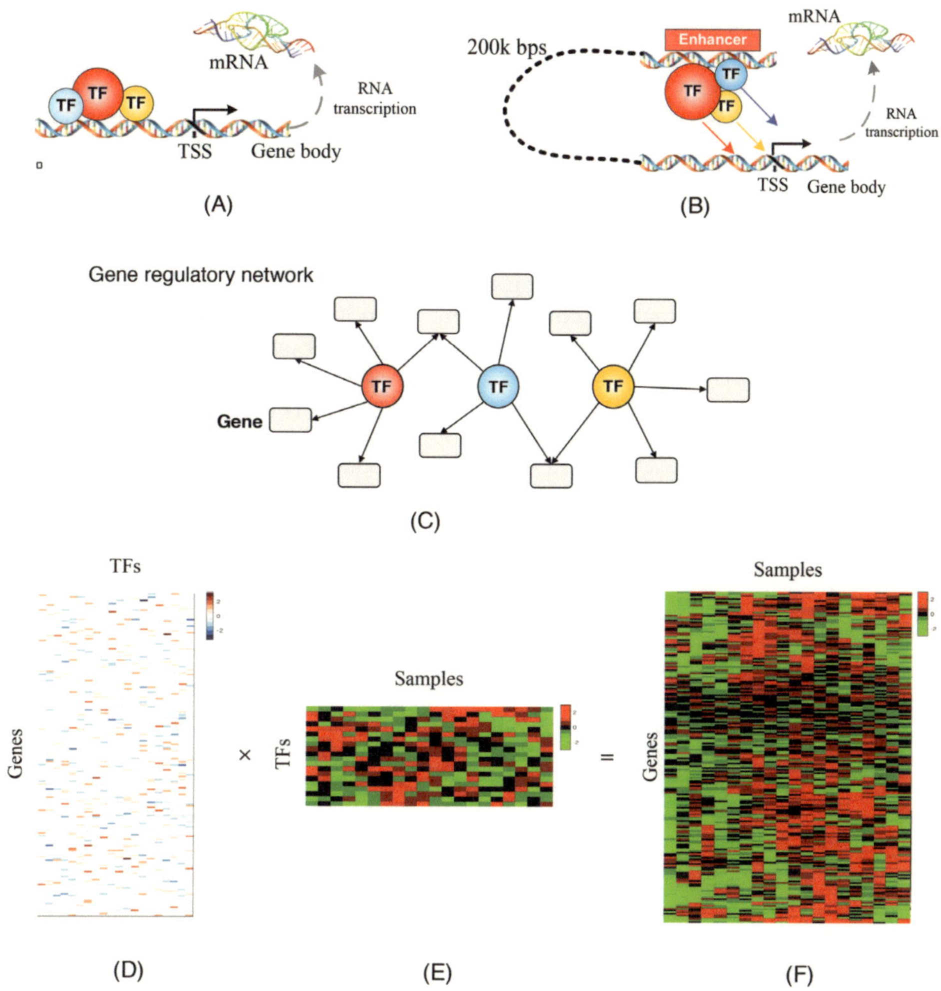

Figure 1.
Illustration of gene regulation: (A) transcription factor (TF)-gene regulation through proximal promoter regions; (B) TF-gene regulation through distal enhancer regions; (C) a gene regulatory network (GRN) including TFs, genes, and their interactions; (D) regulatory effects of TFs on individual genes with "red" as activation, "blue" as depression, and "white" as no regulatory effects; (E) a heatmap of TF protein activities across biological samples of multiple conditions with "red" as enhanced activity, "green" as reduced activity, "black" as no activity; and (F) a heatmap of gene expression across multiple samples, with "red" as up-regulated, "green" as down-regulated, "back" as no change.

TFs are not random [14, 15]. Some TFs tend to co-bind at the same regions more often than with others, i.e. MYC and MAX. One TF can regulate multiple genes, and a target gene can also be regulated by multiple TFs considering the existence of CRMs (**Figure 1C**). For each specific TF-gene interaction in **Figure 1C**, its regulatory effect can be either positive (activating gene expression) or negative (depressing gene expression), as shown in **Figure 1D**. The protein activities of TFs are therefore connected to the dynamic changes of gene expression across multiple samples [13]. To accurately identify GRNs, we need quantitative measures of all types of signals in **Figure 1D–F**. However, due to technical limitations, we can obtain good quality measurements of gene expression, binary measurements (existence or not) of individual TF-gene interactions yet with a high false positive rate, but no measurements of TF activities. To infer GRNs, we must jointly estimate TF activities, TF-gene regulation strengths, and CRMs (TF associations) given gene expression observations.

3. Bayesian inference

Bayesian inference is particularly suitable for inferring GRN as it will learn a posterior distribution for each variable, with a high tolerance on the noise existing in the gene expression data or caused by non-perfect prior assumptions.

3.1 A hierarchical Bayesian model

Given gene expression data under multiple biological samples (conditions), we focus on the expression variation of each gene from its baseline expression because such variation reflects the effects of condition changes. For a specific disease, only genes showing significant expression changes between disease cells and normal cells are interesting candidates. Thus, for gene n, we calculate the log fold change of gene expression under each sample (1,2,3,...,M) to that of baseline condition (0). To model gene expression data of hundreds of genes in the same framework, for genen, we normalize its M log fold change values (indexed by m) to values with 0-mean and 1-standard deviation, denoted by $y_{n,m}$. Then, a linear model is applied to modeling $y_{n,m}$ as follows [16, 17]:

$$y_{n,m} = \sum_{t} a_{n,t} b_{n,t} x_{t,m} + \varepsilon_n, \tag{1}$$

where variable $a_{n,t}$ denotes the regulation strength of TF t on gene n; $b_{n,t}$ is a binary variable denoting the regulation occurrence of TF t on gene n; TF protein activity variable $x_{t,m}$ under condition (sample) m directly connects to gene expression $y_{n,m}$ under the same condition [16]; and the noise variable ε_n denotes inaccuracy of gene expression marvelments.

Given protein-DNA binding measurements of T TFs and N genes (i.e., ENCODE database), we are able to identify TF binding sites at promoter or enhancer regions within 1 Mbps around individual target genes [18]. Each gene can be associated with several regulatory regions, and at each region, there exit a subset of TFs, as a candidate CRM. Then, we may observe multiple candidate CRMs (in total K_n) for gene n, indexed by $c_n = 1, 2, 3, \ldots, k, \ldots, K_n$. Each c_n is associated with a unique set of TF-gene binding events ($b_{c_n,t} = 1$ or $b_{c_n,t} = 0$). We assume c_n a hidden variable controlling how binding variables are associated with each other, with candidate space defined from existing databases.

To estimate the abovementioned variables, we develop a hierarchical Bayesian network to model their internal dependency and associations with gene expression, as shown in **Figure 2**. CRM variable c controls the state of each binding variable b. For $b = 1$, regulation strength a can be either positive or negative denoting gene activation or depression by the binding TF. In the meanwhile, through TF-gene regulation, the protein activities of TFs are directly connected to target gene expression, with ε denoting the measurement noise in gene expression data. With Eq. (1) and **Figure 2**, we aim to estimate all these variables using Bayesian inference, which requires a prior assumption (not necessary to be informative) on the distribution of each variable.

Based on prior binding observations from public database, the candidate space of CRM is known, denoted by **C**. Given a gene expression dataset generated from a specific condition, for gene n, we need to estimate which CRM c_n is regulating its gene expression. As the prior data does not tell which CRM is more likely to be true under a specific condition, we assume a discrete uniform prior on c.

Based on data observation, y has a Gaussian-like distribution with 0-mean and 1-standard deviation. The gene expression noise component ε can be assumed to

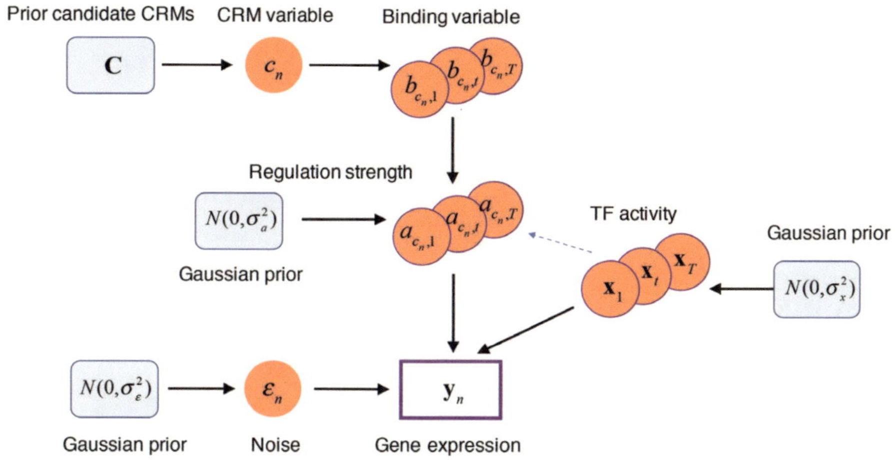

Figure 2.
A hierarchical Bayesian framework for GRN modeling. The number of variables in this framework depends on the numbers of biological samples, TFs, genes, and candidate CRMs. Given gene expression data under different conditions, for the same TF and same gene, their regulatory relationship (variable b) may have very different regulatory strength (variable a). And the TF activity (variable x) can be significantly different as well. Therefore, GRNs are highly context-specific.

follow a 0-mean Gaussian distribution as well, denoted by $N(0, \sigma_\varepsilon^2)$. Although the variance of noise is hard to determine, it should fall in the same scale as gene expression measurements. Therefore, we set $\sigma_\varepsilon^2 = 1$.

The regulation strength variable a is conditional on the state of b (as shown in **Figure 2**): for $b = 0$, we set $a = 0$, denoting the nonexistence of TF-gene regulation; for $b = 1$, a can be either positive or negative so that we assume a 0-mean Gaussian prior on a, as $N\left(0, \sigma_{a,prior}^2\right)$ (the variance $\sigma_{a,prior}^2$ is a hyperparameter). As GRN is a sparse network, most a values would be 0.

We model TF activity x under multiple biological samples using Gaussian random processes. As baseline expression is largely removed from gene expression data during the data normalization process, ideally the baseline activity of each TF is 0. In each sample, x can be either enhanced or reduced with respect to its baseline activity. Thus, we assume a 0-mean Gaussian prior for x, as $N\left(0, \sigma_{x,prior}^2\right)$ (the variance $\sigma_{x,prior}^2$ is also a hyperparameter).

Regarding hyperparameters of the prior mean and variance for a or x, a benefit of assuming 0-mean prior is to control model overfitting. Only when the posterior distribution has a significant non-zero mean value that we will accept that estimation. It is hard to determine the scale of variable values without direct measurements. A conservative way is to assume non-informative prior on them and let the algorithm determine the final posterior distribution, although the non-informative prior will lead to a stickier chain and a posterior with potential multiple modes. Exploring such a posterior is certainly more challenging than exploring a well-behaved unimodal posterior. However, there is really no need to trouble with this multimodal posterior on a or x, as the inferential values of the whole framework are: the discrete posterior distributions of CRMs. For each gene, the posterior distribution of CRMs learned from a data reveals which CRM(s) are regulating this gene. If there are more than one mode in the CRM posterior distribution, this gene will be associated with two or three CRMs. This is quite common in gene regulatory networks as one gene can be regulated by CRMs at multiple regulatory regions simultaneously. $\sigma_{a,prior}^2$ and $\sigma_{x,prior}^2$ should be significantly larger than the variance of gene expression data to allow a "large" space for the algorithm to generate posterior distributions. As y is already normalized with variance of 1, we set $\sigma^2_{a,prior} = 10$ and $\sigma^2_{x,prior} = 100$.

Then, the problem of GRN inference is Bayesian formed as estimating posterior probabilistic distributions of $\mathbf{A} = \{a_{c_n,t}\}$, $\mathbf{B} = \{b_{c_n,t} | b_{c_n,t} = 0 \text{ or } 1\}$, and $\mathbf{X} = \{x_{t,m}\}$ given $\mathbf{Y} = \{y_{n,m}\}$. Considering the dependence relationship of all variables in **Figure 2**, we define a joint posterior probability as follow:

$$\begin{aligned} P(\mathbf{A}, \mathbf{B}, \mathbf{X}|\mathbf{Y}) &\propto P(\mathbf{Y}|\mathbf{A}, \mathbf{B}, \mathbf{X}) \times P(\mathbf{A}) \times P(\mathbf{C}) \times P(\mathbf{X}) \\ &\propto \prod_n \prod_m (\sigma_\varepsilon^{-1}) \exp\left(-\frac{1}{2\sigma_\varepsilon^2}\left(y_{n,m} - \sum_t a_{c_n,t} b_{c_n,t} x_{t,m}\right)^2\right) \\ &\times \prod_n \prod_t \left(\sigma_{a,prior}^{-1}\right) \exp\left(-\frac{a_{c_n,t}^2}{2\sigma_{a,prior}^2}\right) \\ &\times \prod_n \prod_{c_n} \frac{1}{K_n} \\ &\times \prod_t \prod_m \left(\sigma_{x,prior}^{-1}\right) \exp\left(-\frac{x_{t,m}^2}{2\sigma_{x,prior}^2}\right). \end{aligned} \tag{2}$$

Estimating the joint distribution of above-mentioned variables is difficult. Alternatively, we can approximate the joint posterior distribution by estimating the marginal distribution of each variable. To do that, we iteratively calculate each variable's conditional probability and perform Bayesian estimation using Gibbs sampling. The advantage of using Gibbs sampling is that it is theoretically guaranteed to converge to the posterior distribution [2, 19–21].

3.2 Gibbs sampling

We first sample TF activity variable $x_{t,m}$ for the TF t and sample m, according to its conditional probability (based on Eq. (2)) as follows (**Figure 3**):

$$P(x_{t,m}|\mathbf{Y}, \mathbf{A}, \mathbf{B}) \propto \prod_n \exp\left(-\frac{1}{2\sigma_\varepsilon^2}\left(y_{n,m} - \sum_t a_{c_n,t} b_{c_n,t} x_{t,m}\right)^2 - \frac{x_{t,m}^2}{2\sigma_{x,prior}^2}\right). \tag{3}$$

$P(x_{t,m}|\mathbf{Y}, \mathbf{A}, \mathbf{B})$ is a Gaussian distribution with mean and variance as follows:

$$\begin{cases} \mu_x = \dfrac{\sigma_{x,prior}^2 \sum_n \left(y_{n,m} - \sum_{j \neq t} a_{c_n,j} b_{c_n,j} x_{j,m}\right) a_{c_n,t} b_{c_n,t}}{\sigma_{x,prior}^2 \sum_n a_{c_n,t}^2 b_{c_n,t}^2 + \sigma_\varepsilon^2 N} \\ \sigma_x^2 = \dfrac{\sigma_\varepsilon^2 N \sigma_{x,prior}^2}{\sigma_{x,prior}^2 \sum_n a_{c_n,t}^2 b_{c_n,t}^2 + \sigma_\varepsilon^2 N} \end{cases} \tag{4}$$

As shown in Eq. (4), the estimation of distribution of $x_{t,m}$ is conditional on other TF activities $x_{j,m} (j \neq t)$. Therefore, we iteratively sample $x_{t,m}$ as $x_{t,m}|x_{j,m} (j \neq t)$ one

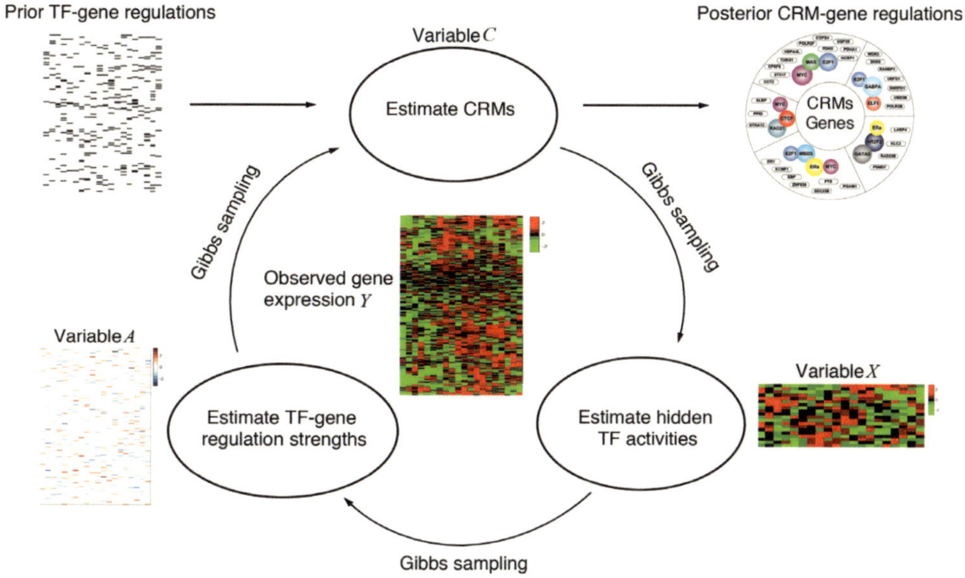

Figure 3.
Gibbs sampling of CRMs, TF activities, and regulation strengths with prior TF-gene regulation and gene expression observations as input.

by one for $t = 1 \sim T$ according to each individual posterior Gaussian distribution $N(\mu_x, \sigma_x^2)$.

Secondly, for gene n, for each $b_{c_n,t} = 1$, we estimate the associated regulation strength $a_{c_n,t}$ according to the following conditional probability:

$$P(a_{c_n,t}|\mathbf{Y}, \mathbf{X}, \mathbf{B}) \propto \prod_m \exp\left(-\frac{1}{2\sigma_\varepsilon^2}\left(y_{n,m} - \sum_t a_{c_n,t} x_{t,m}\right)^2 - \frac{a_{c_n,t}^2}{2\sigma_{a,prior}^2}\right). \quad (5)$$

$P(a_{c_n,t}|\mathbf{Y}, \mathbf{X}, \mathbf{B})$ is a Gaussian distribution, too, with mean and variance calculated as follows:

$$\begin{cases} \mu_a = \dfrac{\sigma_{a,prior}^2 \sum_m \left(y_{n,m} - \sum_{j \neq t} a_{c_n,j} x_{j,m}\right) x_{t,m}}{\sigma_{a,prior}^2 \sum_m x_{t,m}^2 + M\sigma_\varepsilon^2} \\ \sigma_a^2 = \dfrac{\sigma_{a,prior}^2 M \sigma_\varepsilon^2}{\sigma_{a,prior}^2 \sum_m x_{t,m}^2 + \sigma_\varepsilon^2 M} \end{cases} \quad (6)$$

Similar to the estimation process of TF activity variables, the posterior distribution of each $a_{c_n,t}$ also depends on the values of the other $a_{c_n,j} (j \neq t)$. Thus, we iteratively sample $a_{c_n,t}$ for TFs in module c_n one by one according to each individual posterior Gaussian distribution $N(\mu_a, \sigma_a^2)$.

Finally, with sampled TF activity and regulation strength variables, we sample CRM variable c_n for the gene n. It is hard to assume a prior probabilistic distribution shape on the joint distribution of multiple binding variables in c_n. In practice, c_n has a finite number of states as K_n. Therefore we can directly calculate a discrete discrete conditional probability for each $c_n = k$ as follows:

$$P(c_n|\mathbf{y}_n, \mathbf{A}, \mathbf{X}) \propto \prod_t P(\mathbf{y}_n|a_{c_n,t}, \mathbf{x}_t)P(a_{c_n,t}|c_n)P(\mathbf{x}_t)$$
$$\propto \prod_t \exp\left(-\frac{1}{2\sigma_\varepsilon^2}\sum_m \left(y_{n,m} - \sum_t a_{c_n,t}b_{c_n,t}x_{t,m}\right)^2 - \frac{a_{c_n,t}^2}{2\sigma_{a,prior}^2} - \frac{\sum_m x_{m,t}^2}{2\sigma_{x,prior}^2}\right) \tag{7}$$

After calculating Eq. (7) for all possible values of c_n, we sample one value according to the following discrete probability density function:

$$p(c_n = k) = \frac{P(c_n = k|\mathbf{y}_n, \mathbf{X}, \mathbf{A})}{\sum_p P(c_n = p|\mathbf{y}_n, \mathbf{X}, \mathbf{A})} \tag{8}$$

After sampling TFA, TF-gene regulation strength, and cis-regulatory module variables for all N genes, we update binding states in matrix **B** according to the sampled CRMs for individual genes and start the next round of sampling.

Convergence of Gibbs sampling can be monitored based on the ratio (R) of within-variance and between-variance using multiple sequences with different initial states [22]. In each application, we ran five sequences of sampling in parallel. In the i-th round of sampling, for each variable we calculated the within-variance using samples from 1 to i in each sequence and then take the mean value of variances from five sequences. In the meanwhile, we calculate the between-variance of the same variable using its sampled values in the i-th round but from five sequences. For each catalog of variables, the distribution of ratio (R) between within-variance and between-variance is used to monitor the overall sampling convergence. When the sampler converges, values of R would be around "1." We, respectively, monitor the sampling convergence for regulation strengths and TF activities. Once both of them converge, we start to accumulate samples on TF-gene binding variables. As each TF-gene binding variable is binary, its sampling frequency represents the posterior probability of binding occurrence. In the meanwhile, for each gene, a discrete posterior probability distribution of all associated candidate CRMs is inferred, the mode of which reveals the most likely regulatory region associated with current gene.

4. Inferring GRNs for breast cancer

4.1 Application to in vitro breast cancer cell line data

We first applied the hierarchical Bayesian model to gene expression data measured from in vitro breast cancer cell lines. We chose to use cell line data mainly because such data is usually clean and good for validating computational models. Here, we carefully selected two public available breast cancer cell line datasets measured independently but under the same condition (downloadable from the GEO database https://www.ncbi.nlm.nih.gov/geo/, with accession number GSE62789 for Data #1 and accession number GSE51403 for Data #2, both treated by 24 hours of 17b-estradiol (E2) to stimulate breast cancer cells proliferation). The similarity between the two inferred GRNs can be used to evaluate the robustness of GRN inference methods.

For prior TF-gene collection, we checked the ENCODE database (https://www.encodeproject.org/) and selected genome-wide binding profiles of 39 TFs, measured from the same breast cancer cell line. We collected candidate binding events by examining TF binding signals at promoter and distantly associated enhancers associated with each gene. In total we collected 2,319 candidate TF-gene interactions (**Figure 4A**) between 39 TFs and 275 genes, whose gene expression is consistently upregulated in both datasets when breast cancer cells are stimulated to fast proliferate (**Figure 4B** and **C**). We, respectively, applied the hierarchical Bayesian model to the two gene expression datasets with the same prior settings. To monitor the convergence of the sampling process, we ran five sequences with different initial states and sampled 1000 times in each. As shown in **Figure 4C** and **D** (for Data #1), after 100 rounds of sampling, the model started to converge. The sampling frequency on each TF-gene interaction was calculated as the posterior probabilistic weight. We extracted top ~500 most confident TF-gene interactions as the final GRN estimation for each data set and then focused on common interactions between two relevant GRNs.

Here, we specifically compared our approach with three competing methods (COGRIM [20], LASSO [23], and NARROMI [24]). COGRIM was a Bayesian inference approach without modeling on CRMs. It treated individual TF-gene binding events independently. Although such an assumption lowered the model complexity, it made the model less robust against the inaccuracy in the TF-gene binding prior. Moreover, for the TF activity, COGRIM simply treated it as an observed value by directly using TF mRNA expression. Although ideally the variation of mRNA transcription is proportional to the activity change of mRNA-translated protein, currently this correlation is very low in most studies using gene expression. These inaccurate assumptions brought a lot of uncertainty to modeling gene expression data. LASSO used a linear regression model to integrate prior TF-gene interactions and gene expression data and predicted one value for each TF-gene interaction. The NARROMI approach inferred GRNs using gene expression data only without any prior on TF-gene interactions, and also, it made single-value prediction for each interaction based on the mutual information between gene and TF expression values. Theoretically, the Bayesian approach described in this chapter should be

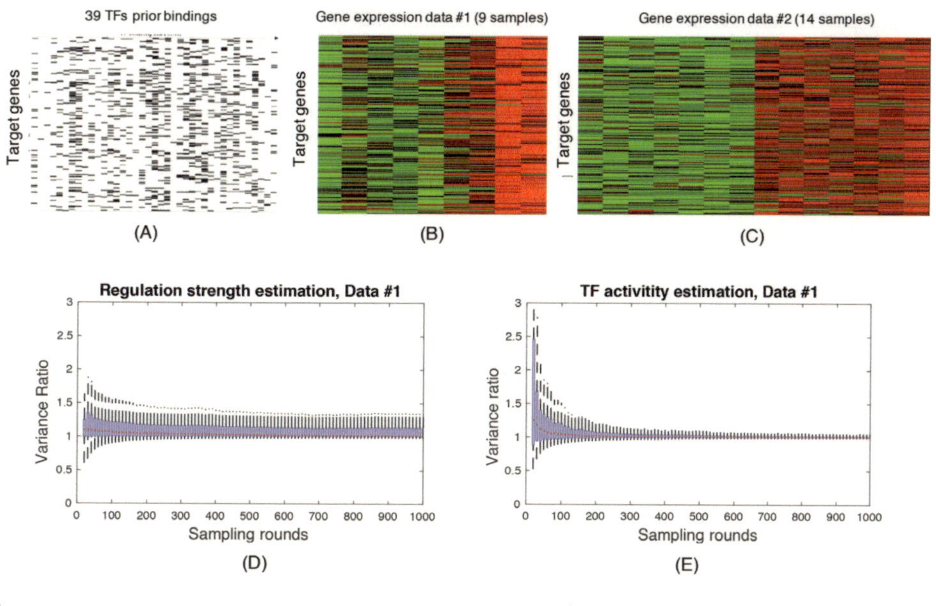

Figure 4.
Input breast cancer cell line data for GRN inference: (A) prior TF-gene interactions ("black" denotes binding occurrence); (B) heatmap of time-course gene expression data; (C) heatmap of steady-state gene expression, all data are from the same breast cancer cell line; (D) convergence of regulatory strength estimation using time-course gene expression data; and (E) convergence of TF activity estimation using time-course gene expression data.

more robust to identify GRNs. We applied the four competing methods to the above two datasets. Indeed, GRNs identified using our Bayesian model were more consistent between two related datasets (**Table 1**).

By analyzing the common 306 TF-gene interactions in **Table 1**, we identified two functional CRMs. The first CRM had five TFs including POL2A, TDRD3, MYC, MAX, and E2F1 (**Figure 5A**). The activities of these TFs, as inferred from both datasets, were shown in **Figure 5B** and **C**, respectively. In total there were 100 genes regulated by this module, and 60 of them were associated with breast cancer through literature survey (selected genes shown in **Figure 5D**). The second CRM had six TFs including ELF1, JUND, JUN, FOXA1, CTCF, and HDAC1. In total, there were 89 genes regulated by this module, and 51 of them were associated with breast cancer (selected genes shown in **Figure 5E**). COGRIM identified fewer genes for the first CRM and failed to identify the second CRM. For the other non-Bayesian approaches, as the number of common TF-gene interactions inferred from two

Methods	GRN edges in Data #1	Similarity with other methods	GRN edges in Data #2	Similarity with other methods	Common GRN for Data #1 and #2
Bayesian	500	**0.878*** **	413	**0.822*** **	**306*** **
COGRIM	516	0.798	457	0.696	239
LASSO	565	0.486	510	0.533	74
NARROMI	514	0.519	591	0.516	44

****denotes hypergeometric p-value < 0.001.*

Table 1.
Comparison of methods for robust GRN inference.

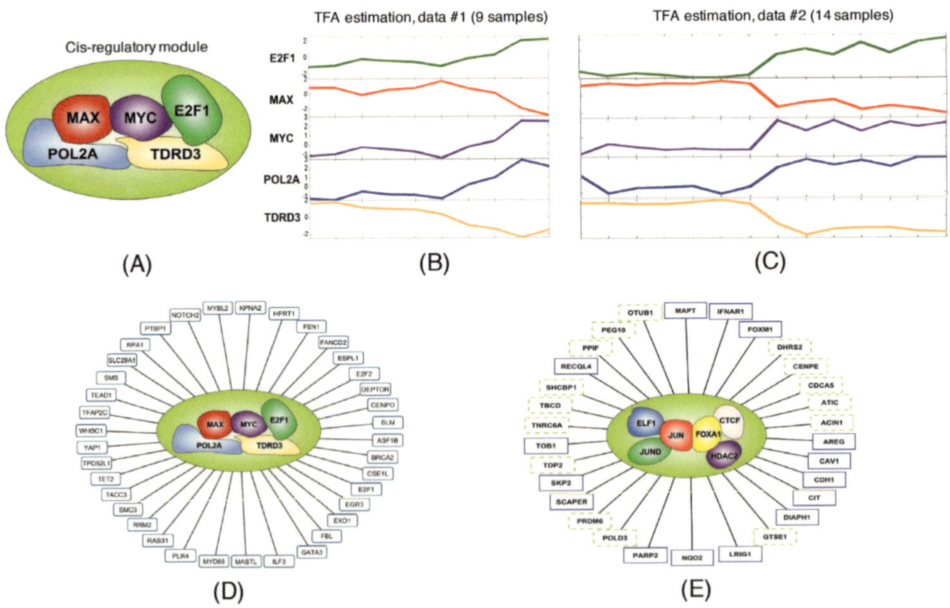

Figure 5.
Key CRMs inferred from breast cancer cell line data: (A) CRM #1 and their TF components; (B) estimated TF activities from Data #1 (time-course); (C) estimated TF activities from Data #2 (steady state); (D) target genes regulated by CRM with MAX, MYC, E2F1, POL2A and TDRD3; (E) target genes regulated by CRM with ELF1, JUND, JUN, FOXA1, CTCF, and HDAC2. Target genes in D and E are associated with breast cancer as supported by literature survey. "Blue" block represents genes showing up in at least two literatures, while "green" block represents genes with one literature support.

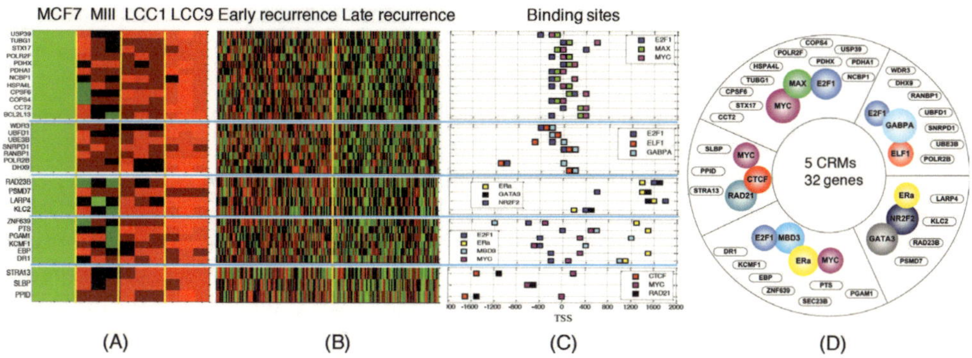

Figure 6.
Breast cancer recurrence-associated GRN: (A) heatmap of gene expression in breast cancer cell lines including MCF7, MIII, LCC1, and LCC9, where "red" represents overexpression and "green" represents lower expression; (B) heatmap of gene expression of breast cancer patients in "Early recurrence" and "Late recurrence" groups, divided by 5-year survival; (C) binding sites of 11 TFs on 32 target genes; and (D) association of 5 CRMs and 32 target genes.

datasets was small, size reduced by over 75%. We did not identify the two key CRMs using either approach.

4.2 Application to breast cancer patient data

We finally applied the Bayesian approach to breast cancer patient data downloaded from the TCGA database (https://portal.gdc.cancer.gov/). Survival time distribution of 93 breast cancer patients treated by *tamoxifen* revealed two modes with 5-year survival as division. Accordingly, we defined an "Early recurrence" group including patients with survival time <5 years and a "Late recurrence" group including patients with survival time longer than 5 years. Differentially expressed genes between two groups (t-test p-value <0.05) were selected for further GRN analysis. It can be seen from **Figure 6B** that the gene expression data of breast cancer patient is quite noisy. To increase the robustness of GRN results, we used another cell line dataset. Specifically, gene expression data was generated from four cell lines including MCF7, MIII, LCC1, and LCC9, with three replicates for each. MCF7 cells were sensitive to *tamoxifen* treatment, while LCC9 cells were drug-resistant. One hypothesis is that breast cancer recurrence is associated with drug resistance. Thus, we expected that the overexpressed genes in the "Early recurrence" group were also overexpressed in LCC9 cells. For 431 genes with such expression pattern in both patient and cell line data, we collected prior TF-gene interactions from 39 TF binding profiles used in previous sections. We, respectively, inferred GRNs using both datasets and identified a common GRN including interactions between 25 proteins and 161 genes. Analysis of this common CRN revealed 5 key CRMs with 11 proteins and 32 target genes highly relevant to breast cancer recurrence (**Figure 6**).

5. Discussion

5.1 Gene regulatory networks in different cell states

Recent technology advance in single-cell gene transcription makes it feasible to study TF-gene regulation during the cell differentiation process [25]. In sections above, across multiple samples, TF-gene interactions are assumed to hold, and the gene expression change is connected to the dynamic variation of TF activities across samples. Yet, at the single-cell level, gene expression measurements are very noisy, whose variation across cells may be partially disconnected from the dynamic changes of TF activities [26]. In that situation, the linear model in Eq. (1) will not work with such gene expression input. Moreover, during the cell differentiation process, in fact we do not have prior knowledge on whether GRNs will hold or change between individual cell states. That means TF-gene interaction change can be another causal factor on gene expression variation across different cell states, too. To model GRNs individually for cell states, we need to define more binding variables, which will definitely make the estimation process more complex.

Those cell state-specific GRNs will uncover the regulatory mechanism that drives cell differentiation. This would be particularly useful for cancer treatment. If any regulation changes at a very early cell state eventually lead to cancer cell fast proliferation, we can engineeringly target those TFs, binding regions, or genes for cancer prevention. Currently inference of cell-state-specific GRN is either through enrichment analysis of TF binding signals in each cell state [27] or regression modeling of gene expression using the matched measurements of regulatory region activities [28]. When the single-cell expression measurements become more accurate, we hope the connection between gene expression and TF activities still holds. Then, the model in Eq. (1) with proper improvement can be used to infer cell-state-specific GRNs.

5.2 Bayesian neural network

Although theoretically there is no upper limit on the number of model parameters in the Bayesian framework (**Figure 2**), the more variables we have, the slower the convergence will be. Moreover, given a complex network with many states, the dependence of different variables will be hard to model, and the estimation process is more easily to stuck into a local state. In recent years, neural network is widely applied to variable estimation in complex systems. Neural network is an end-to-end system that mimics the human brain and tries to learn complex representation within the dataset to provide an output. Similar to conventional machine learning, deep neural networks make a single-value prediction for each model parameter, without measuring uncertainty. That means the model performance relies heavily on the prediction accuracy, and even one overconfident decision can result in a big problem. A Bayesian approach to neural networks can naturally solve this problem by learning a distribution accounting for the uncertainty in parameter estimates [29].

Unlike Bayesian inference discussed in previous sections, inferring model posterior in a Bayesian neural network is much more difficult as there are many parameters to estimate in neural networks. Direct inference of variable posterior distribution is hard so that approximations to the posterior are often used, i.e., the variational inference. The posterior can be modelled using a simple variational distribution such as a Gaussian distribution, and the distribution's parameters are fitted to approximate the true posterior as close as possible by minimizing the Kullback-Leibler divergence between this simple variational distribution and the true posterior. In earlier sections, we have demonstrated that modeling variables in GRN using Gaussian distribution provided robust performance. To infer large-scale GRN with thousands of genes and hundreds of TFs, Bayesian neural network can be a solution in which posterior distributions of all variables can be approximated by Gaussian distribution.

6. Conclusion

In this chapter, we mathematically illustrated how Bayesian inference can be used to infer gene regulatory networks. Using several breast cancer-specific datasets, we demonstrated the effectiveness of Bayesian network modeling in biological meaningful signal discovery, in comparison with methods of linear regression. Potentially, Bayesian inference can be used to infer dynamic GRN during cell differentiation using new types of gene expression data. For very large-scale GRN inference in complex systems, the big number of variables may degrade conventional Bayesian inference performance. Bayesian neural networks using variational inference can be a good solution.

Acknowledgements

Funding for open access charge: Virginia Tech's Open Access Subvention Found (VT OASF).

Author details

Xi Chen[1,2]* and Jianhua Xuan[1]

1 Bradley Department of Electrical and Computer Engineering, Virginia Tech, Arlington, VA, USA

2 Center for Computational Biology, Flatiron Institute, New York, NY, USA

*Address all correspondence to: xichen86@vt.edu

References

[1] Schuster SC. Next-generation sequencing transforms today's biology. Nature Methods. 2008;**5**(1):16-18

[2] Chen X et al. CRNET: An efficient sampling approach to infer functional regulatory networks by integrating large-scale ChIP-seq and time-course RNA-seq data. Bioinformatics. 2018; **34**(10):1733-1740

[3] Barabasi AL, Oltvai ZN. Network biology: Understanding the cell's functional organization. Nature Reviews. Genetics. 2004;**5**(2):101-113

[4] Blais A, Dynlacht BD. Constructing transcriptional regulatory networks. Genes & Development. 2005;**19**(13): 1499-1511

[5] van 't Veer LJ et al. Gene expression profiling predicts clinical outcome of breast cancer. Nature. 2002;**415**(6871): 530-536

[6] Li W, Notani D, Rosenfeld MG. Enhancers as non-coding RNA transcription units: Recent insights and future perspectives. Nature Reviews. Genetics. 2016;**17**(4):207-223

[7] Bock C, Lengauer T. Computational epigenetics. Bioinformatics. 2008;**24**(1): 1-10

[8] Landt SG et al. ChIP-seq guidelines and practices of the ENCODE and modENCODE consortia. Genome Research. 2012;**22**(9):1813-1831

[9] Quackenbush J. Microarray data normalization and transformation. Nature Genetics. 2002;**32**(Suppl): 496-501

[10] Wang Z, Gerstein M, Snyder M. RNA-Seq: A revolutionary tool for transcriptomics. Nature Reviews. Genetics. 2009;**10**(1):57-63

[11] Riethoven JJ. Regulatory regions in DNA: Promoters, enhancers, silencers, and insulators. Methods in Molecular Biology. 2010;**674**:33-42

[12] Chen X, Xuan J, Shi X, Shajahan-Haq AN, Hilakivi-Clarke L, Clarke R. A novel statistical approach to identify co-regulatory gene modules. In: 2013 IEEE International Conference on Bioinformatics and Biomedicine; 2013. pp. 16-18

[13] Spitz F, Furlong EE. Transcription factors: From enhancer binding to developmental control. Nature Reviews. Genetics. 2012;**13**(9):613-626

[14] Wang J et al. Sequence features and chromatin structure around the genomic regions bound by 119 human transcription factors. Genome Research. 2012;**22**(9):1798-1812

[15] Chen X, Shi X, Shajahan-Haq AN, Hilakivi-Clarke L, Clarke R, Xuan J. Statistical identification of co-regulatory gene modules using multiple ChIP-seq experiments. In: Presented at the International Conference on Bioinformatics Models, Methods and Algorithms (Bioinformatics); 2014

[16] Liao JC, Boscolo R, Yang YL, Tran LM, Sabatti C, Roychowdhury VP. Network component analysis: Reconstruction of regulatory signals in biological systems. Proceedings of the National Academy of Sciences of the United States of America. 2003;**100**(26): 15522-15527

[17] Chen X, Xuan J, Wang C, Shajahan AN, Riggins RB, Clarke R. Reconstruction of transcriptional regulatory networks by stability-based network component analysis. IEEE/ACM Transactions on Computational Biology and Bioinformatics. 2013;**10**(6): 1347-1358

[18] Chen X et al. ChIP-BIT: Bayesian inference of target genes using a novel joint probabilistic model of ChIP-seq profiles. Nucleic Acids Research. 2016; **44**(7):e65

[19] Sabatti C, James GM. Bayesian sparse hidden components analysis for transcription regulation networks. Bioinformatics. 2006;**22**(6):739-746

[20] Chen G, Jensen ST, Stoeckert CJ Jr. Clustering of genes into regulons using integrated modeling-COGRIM. Genome Biology. 2007;**8**(1):R4

[21] Shi X et al. mAPC-GibbsOS: An integrated approach for robust identification of gene regulatory networks. BMC Systems Biology. 2013;**7** (Suppl 5):S4

[22] Gelman A, Rubin DB. Inference from iterative simulation using multiple sequences. Statistical Science. 1992;7(4): 457-472

[23] Qin J, Hu Y, Xu F, Yalamanchili HK, Wang J. Inferring gene regulatory networks by integrating ChIP-seq/chip and transcriptome data via LASSO-type regularization methods. Methods. 2014; **67**(3):294-303

[24] Zhang X et al. NARROMI: A noise and redundancy reduction technique improves accuracy of gene regulatory network inference. Bioinformatics. 2013;**29**(1):106-113

[25] Wagner A, Regev A, Yosef N. Revealing the vectors of cellular identity with single-cell genomics. Nature Biotechnology. 2016;**34**(11):1145-1160

[26] Raj A, van Oudenaarden A. Nature, nurture, or chance: Stochastic gene expression and its consequences. Cell. 2008;**135**(2):216-226

[27] Aibar S et al. SCENIC: Single-cell regulatory network inference and clustering. Nature Methods. 2017; **14**(11):10831-11086

[28] Cao J et al. Joint profiling of chromatin accessibility and gene expression in thousands of single cells. Science. 2018;**361**(6409):1380-1385

[29] Crucianu M, Bone R, de Beauville JPA. Bayesian learning for recurrent neural networks. Neurocomputing. 2001;**36**:235-242

Chapter 6

Patient Bayesian Inference: Cloud-Based Healthcare Data Analysis Using Constraint-Based Adaptive Boost Algorithm

Shahid Naseem

Abstract

Cloud-based healthcare data are a form of distributed data over the internet. The internet has become the most vulnerable part of critical healthcare infrastructures. Healthcare data are considered to be sensitive information, which can reveal a lot about a patient. For healthcare data, apart from confidentiality, privacy and protection of data are very sensitive issues. Proactive measures such as early warning are required to reduce the risk of patient's data violation. This chapter investigates the ability of Patient Bayesian Inference (PBI) for network scenario analysis with violation of patient data to produce early warning. The Bayesian inference allows modeling the uncertainties that come with the problem of dealing with missing data, allows integrating data from remote nodes, and explicitly indicates dependence and independence. The use of constraint-based adaptive boost algorithm can demonstrate the patient's Bayesian inference performance in the real-world datasets from healthcare data.

Keywords: Bayesian inference, healthcare, constraint-based learning, explicitly, adaptive

1. Introduction

Healthcare data have always been considered to be sensitive information, which can reveal a lot about a patient. This is why medical confidentiality prohibits a medical professional to disclose information about a patient's case. If a physician does not have accurate information on a patient's health, it may lead to an inaccurate diagnosis and improper treatment. Data concerning health mean personal data related to the physical or mental health of patients, including the provision of healthcare, which are real information about patient's health. Sensitive data concerning health require additional protection as it can go to the core of a human being. Healthcare data come within a person's most intimate sphere. Unauthorized disclosure may lead to various forms of discrimination and violation of fundamental rights. The risk of data processing generally does not depend on the contents of the data but on the context in which they are used [1].

The processing of healthcare data is likely to lead violation of individual rights and interests. Patients' data, which are, by their nature, particularly sensitive in relation to fundamental rights and freedoms. Data processing could create significant risk to the patient's rights and freedoms. In principle, processing of sensitive data is prohibited, unless a suitable safeguard method is used to protect the data [2]. Derogating from the prohibition to process special categories of a patient data including health data is allowed with the following cases [3]:

- Explicit consent is given by the data subject.
- Processing is necessary to protect the vital interest of a patient if this patient is physically or legally incapable to give consent, for example, in emergency situations or with minors.
- Processing is necessary in order to provide healthcare if the data are processed by or under the responsibility of a professional subject to the obligation of professional secrecy.

2. Risks in cloud-based healthcare data

Cloud computing has many risks related to data confidentiality and data security. The data stored in the cloud are highly confidential, such as patient records. Most of time, data being stored or processed in cloud are in large numbers, and the cloud servers sometimes become lazy because of the computation that affects correctness of final result. Therefore, the computation has to be made transparent. Healthcare data mainly contain of large media files such as X-ray, CT scans, radiology, and other type of images and videos. Such files are called as the Electronic Health Records that are stored in distributed storage. Possibly, this patient perception is fueled by the fact that healthcare data may be disclosure to unauthorized person [4].

In order to secure the patient's sensitive data from unauthorized access, an appropriate encryption standard must be applied to data stored in cloud. This sensitive information is most confidential and needs to be protected. To put everything in the cloud in an unencrypted is a big risk. Over the past four decades, a lot of efforts have been put into developing healthcare information security systems. There is a great variety of commercially available programs to assist clinicians with diagnosis, decision making, pattern recognition, medical reasoning, filtering, and so on for general and very specialized domain applications. If a healthcare system is not secured, an adversary could read, modify, and inject messages into the network. Such incorrect information, even when not for nefarious reasons, can lead to serious consequences for patients and for safe services such as remote healthcare monitoring due to using heterogeneous devices that use a variety of communication rules. Most of the rules that are designed for cloud-based communication cannot be directly applied in the cloud-based healthcare network. In cloud-based healthcare system, remote nodes have limited computation, processing, and communication rights [5].

The existing techniques for healthcare data include pseudo copulation (replacing the most identifying fields in a data) and encryption (encoding the data in such a way that only authorized remote institutions can access it). The existing safeguards are referred to as medical confidentiality or doctor-patient privilege, which prohibit a medical professional to disclose information about a patient's case. This is an important obligation within the medical professional in order to create trust between a doctor and his patient and a trusting environment in which the patient feels comfortable. If a patient cannot trust a physician's discretion, he will not seek medical care altogether or will withhold information during a consultation. If a physician does not have accurate information about a patient's health, this may lead to an inaccurate diagnosis and improper treatment, which may lead to great harm to the patient's health [6].

Figure 1 shows a typical information flows in the healthcare network. Patient information serves as a range of purposes apart from diagnosis and treatment provision. Patient information could be used to improve efficiency within the healthcare system. Patient information could be shared with finance facilitators to justify payment of service rendered. Health service providers may share health information through improved service quality. Furthermore, these providers may share health information through Regional Services to facilitate care services in the regional areas [7].

Credentialing is a vital process for all healthcare systems that must be performed to ensure that those healthcare workers who will be providing the clinical services are qualified to do so. The cloud-based healthcare system is capable to ensure patient safety and deliver an acceptable standard of care. While employing excellent medical staff is vital for success, the healthcare system must have to define the required minimum credentialing and privileging requirements to validate the competency of healthcare providers. In the classical systems, only hospitals used to perform credentialing, but our proposed system has capability to provide all healthcare facilities and also to perform credentialing [8].

In this framework, we classify different modules based on the probability (i.e., trust level) of each provider in violating the patient's data in detail. Honestly, I cannot understand exactly what this statement means. Remote nodes (healthcare physicians, nurses, family members, and other authorized individuals) are different from main modules (patients, health service providers, finance facilitators, regional services, and evaluative decisions), and so it is necessary to make clear remote nodes and modules because the patient Bayesian model only evaluates the trusty of

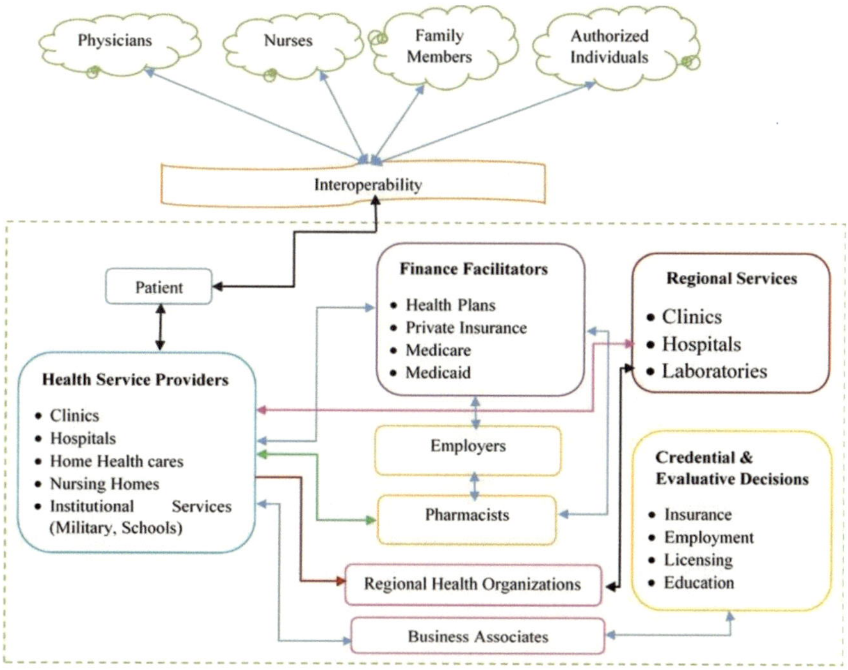

Figure 1.
Cloud-based healthcare system.

remote nodes and whole network based on service level expectation (SLE) as evidence, following the statement "we take advantage of the nature of the Bayesian inference to calculate the probability of wireless communication between the healthcare system and its remote institutions" [9].

3. Problem statement

Healthcare system has become the inspiration for patients' data in terms of wirelessly communication for decision making and logical functionality of the remote institutions such as health physicians, nurses, family members, and authorized individuals. Conventional healthcare systems are using various encryption methods to secure patients' data. Observing the limitations of the existing encryption methods, we take advantage of the nature of the Bayesian inference to calculate the probability of wireless communication between the healthcare system and its remote institutions. The dynamics of the cloud environment requires the healthcare system being able to self-adapt, being aware of its surrounding environment's changing parameters, and being able to create new rules based on past experience. To eliminate the problem of repetition in the cloud environment, the security algorithm must maintain the remote institution limitations and at the same time must provide high level of data protection. Constraint-based adaptive boost algorithm has progressed to an advanced level data analysis for cloud-based healthcare system. The implementation of patient Bayesian Inference for cloud-based healthcare system will be suitable to demonstrate its performance in the real-world patients' datasets. Protection of patient's sensitive data is one of the main obstacles to the growth of cloud computing in the health field because of the need for high level of data integration, interoperability, and sharing among healthcare institutions. It is necessary to create standard guidelines and identify security challenges for improving information security in healthcare system. There are multiple remote institutions (nodes) that have to deal with healthcare data such as healthcare physicians, nurses, family members, and other authorized individuals. Similarly, within healthcare system, there are multiple entities that have to deal with healthcare data such as healthcare providers, hospital administration staff, finance providers, and patients themselves. Cloud services suffer from certain vulnerabilities [10]. By contrast, Bayesian model as an uncertain reasoning tool is more efficient for dynamic trust evaluation. Bayesian inference combined with cloud model and Bayesian network is proposed in this research.

4. Patient Bayesian inference

In cloud-based healthcare systems, patients' electronic data have been widely adapted to improve the quality of patient care and increase the productivity and efficiency of healthcare delivery. In cloud-based systems, patients' data can be helpful to resolve many of the existing problems associated with disease diagnosis and also maintaining the privacy and sensitivity of the patient's medical information. PBI can be beneficial in the healthcare system for tracking fatigue by using multiarmed bandits, which facilitate the healthcare doctors in treatment by dynamically taking more samples from those treatments, which are most likely to be the best. PBI may facilitate the doctors in better understanding the patient's data and make decisions based on it. Because of security in cloud computing, outcomes can be measured in real time, rather than waiting for enough data. Recently, health data privacy has become an important issue in the cloud-based healthcare systems.

As a result, data mining techniques include swapping attribute values and principal component analysis-based techniques, adding random components have gained much more attention in the healthcare data analysis [11].

In healthcare system, PBI is an extremely powerful set of tools that use some knowledge or beliefs to calculate the probability of biomedical and healthcare events, statics, and Service Level Expectation (SLE). PBI can be used for mapping our understanding of a problem and evaluating observed data into a quantitative measure of how certain we are of a particular fact in probabilistic terms, where the probability of a proposition simply represents a degree of belief in the trust of that proposition. PBI can also be used as data mining technique for analyzing network healthcare system variables, virtual assistants, and other variable analytics [12]. PBI uses data and evidence that certain facts are more likely than others. Prior distribution reflects our belief before seeing any data, whereas posterior distributions reflect our belief after we have considered all the evidence.

Cloud-based PBI consists of five main modules: (i) patients; (ii) health service providers; (iii) finance facilitators; (iv) regional services; and (v) traditional and evaluative decisions and four submodules: (a) employers; (b) pharmacists; (c) regional health organizations; and (d) business associates. In this framework, we classify different modules based on the probability (i.e., trust level) of each provider in violating the patient's data. Bayesian rules allow calculating the posterior probability of any information violation events as hypothesis (H) based on a set of historical data (D).

$$P(H|D) = \frac{P(D|H)P(H)}{P(D)} \tag{1}$$

where $P(H|D)$ is the posterior probability of H given knowledge data D; $P(H)$ is the prior probability for H; $P(D|H)$ is the likelihood probability of H given D; and $P(D)$ is the marginal probability that would have happened whether or not H is true. In cloud-based healthcare system, we use Bayes' rule to find the probability function as in Eq. (2):

$$P(\text{SLE, healthcare system, remote nodes}) = P(\text{SLE} \mid \text{healthcare system, remote nodes}) * P(\text{remote nodes} \mid \text{healthcare system}) * P(\text{healthcare system}) \tag{2}$$

SLE is abbreviation of service level expectations. In cloud-based healthcare system, SLE is responsible to provide the quality of services to the remote nodes. It can also be variable that has enough relevance for the service and can be quantitatively and objectively measured. It strengthens the processes to improve the outcomes. In Bayesian Inference, our initial beliefs are represented by the prior distribution P(healthcare system) as shown in **Figure 2**.

In **Figure 2**, remote nodes and healthcare network are hidden variables, and the only observable variable is the SLE metric. An SLE node forecasts how long it should take a share healthcare information to the remote nodes. The SLE itself has two parts: a period of elapsed time and a probability associated with that period (e.g., 38% of healthcare information is shared in 5 min or less, which can also be stated as "5 min with 38% confidence/probability"). However, the healthcare network is a complete system for the variables and their dependencies. Healthcare system can also calculate the services provides to the services provided to the remote nodes like "what is the probability that network successfully passes and the given SLE has failed, P(healthcare system = true | SLE = false), which shows that the sharing of the healthcare information with the remote nodes is not completed within the threshold level. In general, the ultimate purpose of the proposed patient

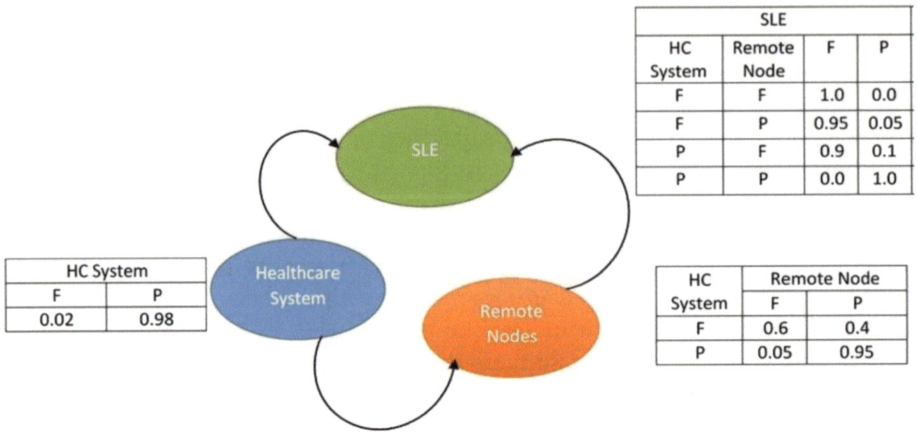

HC System	
F	P
0.02	0.98

SLE			
HC System	Remote Node	F	P
F	F	1.0	0.0
F	P	0.95	0.05
P	F	0.9	0.1
P	P	0.0	1.0

HC System	Remote Node	
	F	P
F	0.6	0.4
P	0.05	0.95

Figure 2.
Communication between healthcare system and remote nodes.

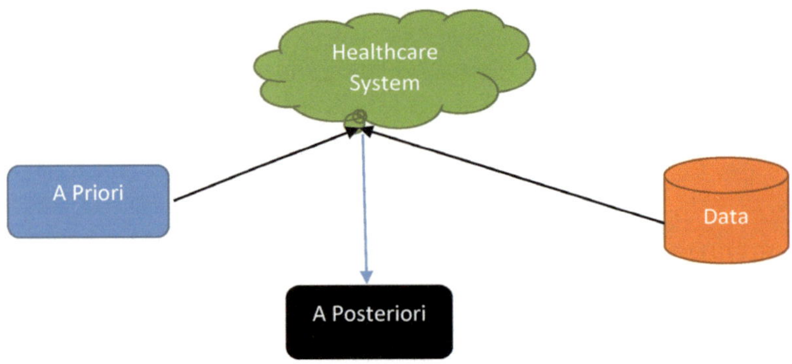

Figure 3.
Bayesian rules.

Bayesian model is to calculate the posterior (conditional) probability of the healthcare system given SLE, P(healthcare system| SLE), which reflects the trusty of the healthcare system. Eq. (3) calculates the posterior probability P(healthcare system | SLE), according to Bayes' rule (**Figure 3**).

It is necessary to choose a probabilistic model represented by Eq. (2) that relates to the random variables and the model parameters associated with it. At the end, Bayes' rules are applied to combine the prior knowledge and the new observed data to find the posterior probability distribution, following Eq. (3).

$$P(\text{healthcare system} \mid \text{SLE}) = \text{Sum of } P(\text{SLE, healthcare system, remote nodes}) \text{ over all values of remote nodes divided by sum of } P(\text{SLE, healthcare system, remote nodes}) \text{ over all values of remote nodes and SLE.} \quad (3)$$

5. Constraint-based adaptive boost algorithm

The constraint-based adaptive boost (CBAB) algorithm is a simple, flexible, and effective classifier [13]. In cloud-based healthcare system, CBAB is used for patient's data analysis. In healthcare system, each patient has different set of records with some common features and unique attributes such as name, age, disease, etc.

Let $D^1_{n(1)}, D^2_{n(2)}, \ldots D^M_{n(M)}$ are the datasets of M patients and the dataset of Pth node contains a total of $n^{(p)}$ samples, and it can be represented as:

$$D^P_{n(p)} = \left\{ \left(X^p_{1(p)}, Y^p_{1(p)}\right), \left(X^p_{2(p)}, Y^p_{2(p)}\right), \ldots \left(X^p_{n(p)}, Y^p_{n(p)}\right) \right\} \quad (4)$$

where X^p_i is the patient's data at P^{th} node and Y^p_i is the decision making that is being consider here. The CBAB algorithm is applied to analyze the health information of each patient for "t" boosting iterations. In the decision making, each unidentified data is represented by (f_n, θ, δ), where f_n represents the selected health parameter, θ is the decision threshold, and δ is the sign of decision, i.e., +1 or −1. CBAB calls a given learning algorithm in a series of loops t = 1, 2, … t. For any health information X_i, the hypothesis h(X_i) means the decision is either +1 or −1. For the Pth patient, $H^P(.)$ is the set of T unidentified data:

$$\left\{ h^{P(1)}(.)\propto^{P(1)}, h^{P(2)}(.)\propto^{P(2)}, \ldots h^{P(T)}(.)\propto^{P(T)} \right\} \quad (5)$$

where $h^{P(t)}$ is the unidentified data at t^{th} iteration and $\propto^{P(t)}$ is the corresponding weight of the unidentified data. For a particular patient's information X_i, the prediction made by the Pth patient can be defined as:

$$H^P(X) = sign\{H^P(X)\} = sign\left\{ \sum_{t=0}^{T} \propto^{P(t)} h^{P(t)}(X_i) \right\} \quad (6)$$

In a cloud-based healthcare system, all the nodes can share a patient's data to each other, and hence each node will receive M-1 information from other nodes. Therefore, each node would integrate specific information. In this way, the healthcare system would value the sensitivity of the patient's information for decision making. To analyze the original patient's data among different nodes in healthcare system is infeasible due to patient's privacy, therefore, we alternate to applying all the other nodes in the training set of P_{th} node, and compare the error rate of each node with the training rate of P_{th} node as shown in Eq.(7):

The node receiving information from any other node might be changed data, hence before using such data, the P^{th}node should select a suitable subset of relevant data based on f_n. For the P^{th}node, the error rate of q^{th}node is given by:

$$\in^{(q)}_P = \frac{1}{n^{(P)}} \left[\sum_{i=1}^{n^{(P)}} I(sign\{H^q(X^P_i) \neq Y^P_i\} \right] \quad (7)$$

where $H^q(.)$ is the selected information patterns from patient's shared data by node q, and I(.) is the indicator function. The training rate of the P^{th}trained node is given by:

$$\in_P = \frac{1}{n^P} \left[\sum_{i=1}^{n^{(P)}} I(sign\{H^P(X^P_i) \neq Y^P_i\} \right] \quad (8)$$

For every node, we compute the difference between $\in^{(q)}_P$ and $\in_P$. If $\left(\in^{(q)}_P - \in_P\right)$ is less than a certain threshold level, then we can assume that the patient's data shared between P^{th} and q^{th} nodes are similar and we can use q^{th} node as trust node for P^{th} node

6. Conclusion

In our research, we have proposed a patient Bayesian Interference for analyzing the healthcare system. The Bayesian Inference is used to model the uncertainties that come with the problems and dealing with missing data and also allow integrating data from remote resources. We have also used the concept of constraint-based adaptive boosting to demonstrate the patient's Bayesian inference performance in the real datasets from healthcare system to remote resources. In the future, we will try to find more accurate ways to protect the patient's data more accurately without compromising on patient's privacy.

Author details

Shahid Naseem
Department of Information Sciences, Division of Science and Technology, University of Education, Lahore, Pakistan

*Address all correspondence to: shahid.naseem@ue.edu.pk

References

[1] Karim A, Abderrahim H, Hayat K. Big healthcare data: Preserving security and privacy. Journal of Big Data. 2018;**5** (1):1-8

[2] Jawwad A, Ali K. Understanding privacy violations in healthcare big data systems. IEEE. 2018;**20**(3): 273-281

[3] O'Mareen S, Richie O, Peter D. Patient consent to publication and data sharing in industry and NIH-funded clinical trials. Springer Nature. 2018; **269**(19):10-25

[4] Hina A, Jawad H, Junaid C, Kashif S, Mehmet A, Jalal M, et al. Risk analysis of cloud sourcing in healthcare and public health industry. IEEE Access.2018; **6**:35-55

[5] Isra's Ahmed S, Ahmed Mousa A. Security and privacy issues in Ehealthcare systems: Towards trusted security. IJACSA. 2016;**7**(9):229-236

[6] Aaron B, William H, Michal M, Abdemour K, Thar B, Casimino A. An investigation into healthcare data patterns. MDPI. 2019;**11**(2):1-23

[7] Faruk A, Aftab A, Haider A, Nur Al Hassan H. A cloud-based healthcare framework for security & patient's data privacy using Wireless Body Area Network. Elsevier. 2014;**34**:511-517

[8] Roshan P, Sandeep S. Credentialing. India: NCBI; 2019

[9] Ella G, Jae S, Amanda D, Wenday S. Averse health events associated with clinical placement: a systematic review. Elsevier: Nurse Education Today. 2019; **76**(1):178-190

[10] Tatiana E, Geboren M. Security and Acceptance of Cloud Computing in Healthcare. Berlin: Der Technischen Universiat; 2015

[11] Alther M, Redday C. Clinical decision support systems. In: Redday C, Aggarwal C, editors. Healthcare Data Analytics. London: Chapman and Hall Press; 2015. pp. 225-260

[12] Rafiqullah S, Sagar R, Anshul S, Nasar Uddin A. Bayesian method for modeling male breast cancer survival data. APJCP. 2014;**15**(2):663-669

[13] Li Y, Bai C, Reddy C. A distributed ensemble approach for mining healthcare data under privacy constraints. Information Sciences. 2016; **330**:245-259

Chapter 7

The Bayesian Posterior Estimators under Six Loss Functions for Unrestricted and Restricted Parameter Spaces

Ying-Ying Zhang

Abstract

In this chapter, we have investigated six loss functions. In particular, the squared error loss function and the weighted squared error loss function that penalize overestimation and underestimation equally are recommended for the unrestricted parameter space $(-\infty, \infty)$; Stein's loss function and the power-power loss function, which penalize gross overestimation and gross underestimation equally, are recommended for the positive restricted parameter space $(0, \infty)$; the power-log loss function and Zhang's loss function, which penalize gross overestimation and gross underestimation equally, are recommended for $(0, 1)$. Among the six Bayesian estimators that minimize the corresponding posterior expected losses (PELs), there exist three strings of inequalities. However, a string of inequalities among the six smallest PELs does not exist. Moreover, we summarize three hierarchical models where the unknown parameter of interest belongs to $(0, \infty)$, that is, the hierarchical normal and inverse gamma model, the hierarchical Poisson and gamma model, and the hierarchical normal and normal-inverse-gamma model. In addition, we summarize two hierarchical models where the unknown parameter of interest belongs to $(0, 1)$, that is, the beta-binomial model and the beta-negative binomial model. For empirical Bayesian analysis of the unknown parameter of interest of the hierarchical models, we use two common methods to obtain the estimators of the hyperparameters, that is, the moment method and the maximum likelihood estimator (MLE) method.

Keywords: Bayesian estimators, power-log loss function, power-power loss function, restricted parameter spaces, Stein's loss function, Zhang's loss function

1. Introduction

In Bayesian analysis, there are four basic elements: the data, the model, the prior, and the loss function. A Bayesian estimator minimizes some posterior expected loss (PEL) function. We confine our interests to six loss functions in this chapter: the squared error loss function (well known), the weighted squared error loss function ([1], p. 78), Stein's loss function [2–10], the power-power loss function [11], the power-log loss function [12], and Zhang's loss function [13]. It is worthy to note that among the six loss functions, the first and second loss functions are defined on $\Theta = (-\infty, \infty)$, and they penalize overestimation and underestimation equally. The third and fourth loss functions are defined on $\Theta = (0, \infty)$, and they penalize gross overestimation and gross underestimation equally, that is, an action a will suffer an infinite loss when it tends to 0 or ∞. The fifth and sixth loss functions are defined on $\Theta = (0, 1)$, and they penalize gross overestimation and gross underestimation equally, that is, an action a will suffer an infinite loss when it tends to 0 or 1.

The squared error loss function and the weighted squared error loss function have been used by many authors for the problem of estimating the variance, σ^2, based on a random sample from a normal distribution with mean μ unknown (see, for instance, [14, 15]). As pointed out by [16], the two loss functions penalize equally for overestimation and underestimation, which is fine for the unrestricted parameter space $\Theta = (-\infty, \infty)$.

For $\Theta = (0, \infty)$, the positive restricted parameter space, where 0 is a natural lower bound and the estimation problem is not symmetric, we should not choose the squared error loss function and the weighted squared error loss function but choose a loss function which can penalize gross overestimation and gross underestimation equally, that is, an action a will suffer an infinite loss when it tends to 0 or ∞. Stein's loss function owns this property, and thus it is recommended for $\Theta = (0, \infty)$ by many researchers (e.g., see [2–10]). Moreover, [11] proposes the power-power loss function which not only penalizes gross overestimation and gross underestimation equally but also has balanced convergence rates or penalties for its argument too large and too small. Therefore, Stein's loss function and the power-power loss function are recommended for $\Theta = (0, \infty)$.

Analogously, for a restricted parameter space $\Theta = (0, 1)$, where 0 and 1 are two natural bounds and the estimation problem is not symmetric, we should not select the squared error loss function and the weighted squared error loss function but select a loss function which can penalize gross overestimation and gross underestimation equally, that is, an action a will suffer an infinite loss when it tends to 0 or 1. It is worthy to note that Stein's loss function and the power-power loss function are also not appropriate in this case. The power-log loss function proposed by [12] has this property. Moreover, they propose six properties for a good loss function on $\Theta = (0, 1)$. Specifically, the power-log loss function is convex in its argument, attains its global minimum at the true unknown parameter, and penalizes gross overestimation and gross underestimation equally. Apart from the six properties, [13] proposes the seventh property, that is, balanced convergence rates or penalties for the argument too large and too small, for a good loss function on $\Theta = (0, 1)$. Therefore, the power-log loss function and Zhang's loss function are recommended for $\Theta = (0, 1)$.

The rest of the chapter is organized as follows. In Section 2, we obtain two Bayesian estimators for $\theta \in \Theta = (-\infty, \infty)$ under the squared error loss function and the weighted squared error loss function. In Section 3, we obtain two Bayesian estimators for $\theta \in \Theta = (0, \infty)$ under Stein's loss function and the power-power loss function. In Section 4, we obtain two Bayesian estimators for $\theta \in \Theta = (0, 1)$ under the power-log loss function and Zhang's loss function. In Section 5, we summarize three strings of inequalities in a theorem. Some conclusions and discussions are provided in Section 6.

2. Bayesian estimation for $\theta \in (-\infty, \infty)$

There are two loss functions which are defined on $\Theta = (-\infty, \infty)$ and penalize overestimation and underestimation equally, that is, the squared error loss function (well known) and the weighted squared error loss function (see [1], p. 78).

2.1 Squared error loss function

The Bayesian estimator under the squared error loss function (well known), $\delta_2^\pi(\boldsymbol{x})$, minimizes the posterior expected squared error loss (PESEL), $\mathrm{E}[L_2(\theta,a)|\boldsymbol{x}]$, that is,

$$\delta_2^\pi(\boldsymbol{x}) = \arg\min_{a\in\mathcal{A}} \mathrm{E}[L_2(\theta,a)|\boldsymbol{x}], \tag{1}$$

where $\mathcal{A}\{a(\boldsymbol{x}) : a(\boldsymbol{x}) \in (-\infty,\infty)\}$ is the action space, $a = a(\boldsymbol{x}) \in (-\infty,\infty)$ is an action (estimator),

$$L_2(\theta,a) = (\theta-a)^2 \tag{2}$$

is the squared error loss function, and $\theta\in(-\infty,\infty)$ is the unknown parameter of interest. The PESEL is easy to obtain (see [16]):

$$PESEL(\pi,a|\boldsymbol{x}) = \mathrm{E}[L_2(\theta,a)|\boldsymbol{x}] = a^2 - 2a\mathrm{E}(\theta|\mathbf{x}) + \mathrm{E}(\theta^2|\boldsymbol{x}). \tag{3}$$

It is found in [16] that

$$\delta_2^\pi(\boldsymbol{x}) = \mathrm{E}(\theta|\boldsymbol{x}) \tag{4}$$

by taking partial derivative of the PESEL with respect to a and setting it to 0.

2.2 Weighted squared error loss function

The Bayesian estimator under the weighted squared error loss function, $\delta_{w2}^\pi(\boldsymbol{x})$, minimizes the posterior expected weighted squared error loss (PEWSEL) (see [1]), $\mathrm{E}[L_{w2}(\theta,a)|\boldsymbol{x}]$, that is,

$$\delta_{w2}^\pi(\boldsymbol{x}) = \arg\min_{a\in\mathcal{A}} \mathrm{E}[L_{w2}(\theta,a)|\boldsymbol{x}], \tag{5}$$

where $\mathcal{A}\{a(\boldsymbol{x}) : a(\boldsymbol{x}) \in (-\infty,\infty)\}$ is the action space, $a = a(\boldsymbol{x}) \in (-\infty,\infty)$ is an action (estimator),

$$L_{w2}(\theta,a) = \frac{1}{\theta^2}(\theta-a)^2 \tag{6}$$

is the weighted squared error loss function, and $\theta\in(-\infty,\infty)$ is the unknown parameter of interest. The PEWSEL is easy to obtain (see [1]):

$$PEWSEL(\pi,a|\boldsymbol{x}) = \mathrm{E}[L_{w2}(\theta,a)|\boldsymbol{x}] = a^2\mathrm{E}\left(\frac{1}{\theta^2}|\boldsymbol{x}\right) - 2a\mathrm{E}\left(\frac{1}{\theta}|\boldsymbol{x}\right) + 1. \tag{7}$$

It is found in [1] that

$$\delta_{w2}^\pi(\boldsymbol{x}) = \frac{\mathrm{E}\left(\frac{1}{\theta}|\boldsymbol{x}\right)}{\mathrm{E}\left(\frac{1}{\theta^2}|\boldsymbol{x}\right)} \tag{8}$$

by taking partial derivative of the PEWSEL with respect to a and setting it to 0.

3. Bayesian estimation for $\theta \in (0,\infty)$

There are many hierarchical models where the parameter of interest is $\theta\in\Theta = (0,\infty)$. As pointed out in the introduction, we should calculate and use the

Bayesian estimator of the parameter θ under Stein's loss function or the power-power loss function because they penalize gross overestimation and gross underestimation equally. We list several such hierarchical models as follows.

Model (a) (hierarchical normal and inverse gamma model). This hierarchical model has been investigated by [10, 16, 17]. Suppose that we observe $X_1, X_2, \ldots, X_n$ from the hierarchical normal and inverse gamma model:

$$\begin{cases} X_i|\theta \overset{\text{iid}}{\sim} N(\mu,\theta), \quad i = 1, 2, \ldots, n, \\ \theta \sim IG(\alpha,\beta), \end{cases} \tag{9}$$

where $-\infty < \mu < \infty$, $\alpha > 0$, and $\beta > 0$ are known constants, θ is the unknown parameter of interest, $N(\mu,\theta)$ is the normal distribution, and $IG(\alpha,\beta)$ is the inverse gamma distribution. It is worthy to note that the problem of finding the Bayesian rule under a conjugate prior is a standard problem and the problem is treated in almost every text on mathematical statistics. The idea of selecting an appropriate prior from the conjugate family was put forward by [18]. Specifically, Bayesian estimation of θ under the prior $IG(\alpha,\beta)$ is studied in Example 4.2.5 (p. 236) of [17] and in Exercise 7.23 (p. 359) of [16]. However, they only calculate the Bayesian estimator with respect to $IG(\alpha,\beta)$ prior under the squared error loss, $\delta_2^\pi(\boldsymbol{x}) = \text{E}(\theta|\boldsymbol{x})$.

Model (b) (hierarchical Poisson and gamma model). This hierarchical model has been investigated by [1, 16, 19, 20]. Suppose that $X_1, X_2, \ldots, X_n$ are observed from the hierarchical Poisson and gamma model:

$$\begin{cases} X_i|\theta \overset{\text{iid}}{\sim} P(\theta), \quad i = 1, 2, \ldots, n, \\ \theta \sim G(\alpha,\beta), \end{cases} \tag{10}$$

where $\alpha > 0$ and $\beta > 0$ are hyperparameters to be determined, $P(\theta)$ is the Poisson distribution with an unknown mean $\theta > 0$, and $G(\alpha,\beta)$ is the gamma distribution with an unknown shape parameter α and an unknown rate parameter β. The gamma prior $G(\alpha,\beta)$ is a conjugate prior for the Poisson model, so that the posterior distribution of θ is also a gamma distribution. The hierarchical Poisson and gamma model (10) has been considered in Exercise 4.32 (p. 196) of [4]. It has been shown that the marginal distribution of X is a negative binomial distribution if α is a positive integer. The Bayesian estimation of θ under the gamma prior is studied in [19] and in Tables 3.3.1 (p. 121) and 4.2.1 (p. 176) of [1]. However, they only calculated the Bayesian posterior estimator of θ under the squared error loss function.

Model (c) (hierarchical normal and normal-inverse-gamma model). This hierarchical model has been investigated by [2, 21, 22]. Let the observations $X_1, X_2, \ldots, X_n$ be from the hierarchical normal and normal-inverse-gamma model:

$$\begin{cases} X_i|(\mu,\theta) \overset{\text{iid}}{\sim} N(\mu,\theta), \quad i = 1, 2, \ldots, n, \\ \mu|\theta \sim N(\mu_0,\theta/\kappa_0), \theta \sim IG\left(v_0/2, v_0\sigma_0^2/2\right), \end{cases} \tag{11}$$

where $-\infty < \mu_0 < \infty$, $\kappa_0 > 0$, $v_0 > 0$, and $\sigma_0 > 0$ are known hyperparameters, $N(\mu,\theta)$ is a normal distribution with an unknown mean μ and an unknown variance θ, $\mu|\theta$ is $N(\mu_0, \ \theta/\kappa_0)$ which is a normal distribution, and θ is $IG\left(v_0/2, v_0\sigma_0^2/2\right)$ which is an inverse gamma distribution. More specifically, with a joint conjugate prior $\pi(\mu,\theta) \sim N - IG\left(\mu_0, \kappa_0, v_0, \sigma_0^2\right)$, which is the normal-inverse-gamma distribution, the posterior distribution of θ was studied in Example 1.5.1 (p. 20) of [21] and Part I (pp. 69–70) of [22]. However, they did not provide any Bayesian posterior estimator of θ. Moreover, the normal distribution with a normal-inverse-gamma prior which assumes that μ is unknown is more realistic than the normal distribution with an inverse gamma prior investigated by [10] which assumes that μ is known.

3.1 Stein's loss function

3.1.1 One-dimensional case

The Bayesian estimator under Stein's loss function, $\delta_s^\pi(\boldsymbol{x})$, minimizes the posterior expected Stein's loss (PESL) (see [1, 10, 16]), $\mathrm{E}[L_s(\theta,a)|\boldsymbol{x}]$, that is,

$$\delta_s^\pi(\boldsymbol{x}) = \arg\min_{a\in\mathcal{A}} \mathrm{E}[L_s(\theta,a)|\boldsymbol{x}], \tag{12}$$

where $\mathcal{A}\{a(\boldsymbol{x}) : a(\boldsymbol{x}) > 0\}$ is the action space, $a = a(\boldsymbol{x}) > 0$ is an action (estimator),

$$L_s(\theta,a) = \frac{a}{\theta} - 1 - \log\frac{a}{\theta} \tag{13}$$

is Stein's loss function, and $\theta > 0$ is the unknown parameter of interest. The PESL is easy to obtain (see [10]):

$$PESL(\pi,a|\boldsymbol{x}) = \mathrm{E}[L_s(\theta,a)|\boldsymbol{x}] = a\mathrm{E}\left(\frac{1}{\theta}|\boldsymbol{x}\right) - 1 - \log a + \mathrm{E}(\log\theta|\boldsymbol{x}). \tag{14}$$

It is found in [10] that

$$\delta_s^\pi(\boldsymbol{x}) = \frac{1}{\mathrm{E}\left(\frac{1}{\theta}|\boldsymbol{x}\right)} \tag{15}$$

by taking partial derivative of the PESL with respect to a and setting it to 0. The PESLs evaluated at the Bayesian estimators are (see [10])

$$\begin{aligned} PESL_s(\pi,\boldsymbol{x}) &= \mathrm{E}[L_s(\theta,a)|\boldsymbol{x}]|_{a=\delta_s^\pi(\boldsymbol{x})}, \\ PESL_2(\pi,\boldsymbol{x}) &= \mathrm{E}[L_s(\theta,a)|\boldsymbol{x}]|_{a=\delta_2^\pi(\boldsymbol{x})}, \end{aligned} \tag{16}$$

where $\delta_2^\pi(\boldsymbol{x}) = \mathrm{E}(\theta|\boldsymbol{x})$ is the Bayesian estimator under the squared error loss function.

For the variance parameter θ of the hierarchical normal and inverse gamma model (9), [10] recommends and analytically calculates the Bayesian estimator:

$$\delta_s^\pi(\boldsymbol{x}) = \frac{1}{\alpha^*\beta^*}, \tag{17}$$

where

$$\alpha^* = \alpha + \frac{n}{2} \text{ and } \beta^* = \left[\frac{1}{\beta} + \frac{1}{2}\sum_{i=1}^{n}(x_i-\mu)^2\right]^{-1}, \tag{18}$$

with respect to $IG(\alpha,\beta)$ prior under Stein's loss function. This estimator minimizes the PESL. [10] also analytically calculates the Bayesian estimator,

$$\delta_2^\pi(\boldsymbol{x}) = \mathrm{E}(\theta|\boldsymbol{x}) = \frac{1}{(\alpha^* - 1)\beta^*}, \tag{19}$$

with respect to $IG(\alpha, \beta)$ prior under the squared error loss, and the corresponding PESL. [10] notes that

$$\mathrm{E}(\log\theta|\boldsymbol{x}) = -\log\beta^* - \psi(\alpha^*), \tag{20}$$

which is essential for the calculation of

$$PESL_s(\pi, \boldsymbol{x}) = \log\alpha^* - \psi(\alpha^*) \tag{21}$$

and

$$PESL_2(\pi, \boldsymbol{x}) = \frac{1}{\alpha^* - 1} + \log(\alpha^* - 1) - \psi(\alpha^*), \tag{22}$$

depends on the digamma function $\psi(\cdot)$. Finally, the numerical simulations exemplify that $PESL_s(\pi, \boldsymbol{x})$ and $PESL_2(\pi, \boldsymbol{x})$ depend only on α and n and do not depend on μ, β, and $\boldsymbol{x}$; the estimators $\delta_s^\pi(\boldsymbol{x})$ are unanimously smaller than the estimators $\delta_2^\pi(\boldsymbol{x})$; and $PESL_s(\pi, \boldsymbol{x})$ are unanimously smaller than $PESL_2(\pi, \boldsymbol{x})$.

For the hierarchical Poisson and gamma model (43), [20] first calculates the posterior distribution of θ, $\pi(\theta|\boldsymbol{x})$, and the marginal pmf of x, $\pi(x)$, in Theorem 1 of their paper. [20] then calculates the Bayesian posterior estimators $\delta_s^\pi(\boldsymbol{x})$ and $\delta_2^\pi(\boldsymbol{x})$, and the PESLs $PESL_s(\pi, \boldsymbol{x})$ and $PESL_2(\pi, \boldsymbol{x})$, and they satisfy two inequalities. After that, the estimators of the hyperparameters of the model (10) by the moment method $\alpha_1(n)$ and $\beta_1(n)$ are summarized in Theorem 2 of their paper. Moreover, the estimators of the hyperparameters of the model (10) by the maximum likelihood estimator (MLE) method $\alpha_2(n)$ and $\beta_2(n)$ are summarized in Theorem 3 of their paper. Finally, the empirical Bayesian estimators of the parameter of the model (10) under Stein's loss function by the moment method and the MLE method are summarized in Theorem 4 of their paper. In numerical simulations of [20], they have illustrated the two inequalities of the Bayesian posterior estimators and the PESLs, the moment estimators and the MLEs are consistent estimators of the hyperparameters, and the goodness of fit of the model to the simulated data. The numerical results indicate that the MLEs are better than the moment estimators when estimating the hyperparameters. Finally, [20] exploits the attendance data on 314 high school juniors from two urban high schools to illustrate their theoretical studies.

For the variance parameter θ of the normal distribution with a normal-inverse-gamma prior (11), [23] recommends and analytically calculates the Bayesian posterior estimator, $\delta_s^\pi(\boldsymbol{x})$, with respect to a conjugate prior $\mu|\theta \sim N(\mu_0, \theta/\kappa_0)$, and $\theta \sim IG(v_0/2, v_0\sigma_0^2/2)$ under Stein's loss function which penalizes gross overestimation and gross underestimation equally. This estimator minimizes the PESL. As comparisons, the Bayesian posterior estimator, $\delta_2^\pi(\boldsymbol{x}) = \mathrm{E}(\theta|\boldsymbol{x})$, with respect to the same conjugate prior under the squared error loss function, and the PESL at $\delta_2^\pi(\mathbf{x})$, are calculated. The calculations of $\delta_s^\pi(\boldsymbol{x})$, $\delta_2^\pi(\boldsymbol{x})$, $PESL_s(\pi, \boldsymbol{x})$, and $PESL_2(\pi, \boldsymbol{x})$ depend only on $\mathrm{E}(\theta|\boldsymbol{x})$, $\mathrm{E}(\theta^{-1}|\boldsymbol{x})$, and $\mathrm{E}(\log\theta|\boldsymbol{x})$. The numerical simulations exemplify their theoretical studies that the PESLs depend only on v_0 and n, but do not depend on μ_0, κ_0, σ_0, and especially $\mathbf{x}$. The estimators $\delta_2^\pi(\mathbf{x})$ are unanimously larger than the estimators $\delta_s^\pi(\boldsymbol{x})$, and $PESL_2(\pi, \boldsymbol{x})$ are unanimously larger than $PESL_s(\pi, \boldsymbol{x})$. Finally, [23] calculates the Bayesian posterior estimators and the PESLs of the monthly simple returns of the Shanghai Stock Exchange (SSE) Composite Index, which also exemplify the theoretical studies of the two inequalities of the Bayesian posterior estimators and the PESLs.

3.1.2 Multidimensional case

For estimating a covariance matrix which is assumed to be positive definite, many researchers exploit the multidimensional Stein's loss function (e.g., see [2, 8, 24–31]). The multidimensional Stein's loss function (see [2]) is originally defined to estimate the $p \times p$ unknown covariance matrix Σ by $\hat{\Sigma}$ with the loss function:

$$L\left(\Sigma, \hat{\Sigma}\right) = \text{tr}\Sigma^{-1}\hat{\Sigma} - \log\det\Sigma^{-1}\hat{\Sigma} - p. \tag{23}$$

When $p = 1$, the multidimensional Stein's loss function reduces to

$$L_s\left(\sigma^2, a\right) = \frac{a}{\sigma^2} - \log\frac{a}{\sigma^2} - 1, \tag{24}$$

which is in the form of (13), the one-dimensional Stein's loss function.

3.2 Power-power loss function

The Bayesian estimator under the power-power loss function, $\delta_p^\pi(\boldsymbol{x})$, minimizes the posterior expected power-power loss (PEPL) (see [11]), $\text{E}\left[L_p(\theta, a)|\boldsymbol{x}\right]$, that is,

$$\delta_p^\pi(\boldsymbol{x}) = \arg\min_{a \in \mathcal{A}} \text{E}\left[L_p(\theta, a)|\boldsymbol{x}\right], \tag{25}$$

where $\mathcal{A}\{a(\boldsymbol{x}) : a(\boldsymbol{x}) > 0\}$ is the action space, $a = a(\boldsymbol{x}) > 0$ is an action (estimator),

$$L_p(\theta, a) = \frac{a}{\theta} + \frac{\theta}{a} - 2 \tag{26}$$

is the power-power loss function, and $\theta > 0$ is the unknown parameter of interest. The PEPL is easy to obtain (see [11]):

$$PEPL(\pi, a|\boldsymbol{x}) = \text{E}\left[L_p(\theta, a)|\boldsymbol{x}\right] = a\text{E}\left(\frac{1}{\theta}|\boldsymbol{x}\right) + \frac{1}{a}\text{E}(\theta|\boldsymbol{x}) - 2. \tag{27}$$

It is found in [11] that

$$\delta_p^\pi(\boldsymbol{x}) = \sqrt{\frac{\text{E}(\theta|\boldsymbol{x})}{\text{E}\left(\frac{1}{\theta}|\boldsymbol{x}\right)}} \tag{28}$$

by taking partial derivative of the PEPL with respect to a and setting it to 0. The PEPLs evaluated at the Bayesian estimators are (see [11])

$$\begin{aligned} PEPL_p(\pi, \boldsymbol{x}) &= \text{E}\left[L_p(\theta, a)|\boldsymbol{x}\right]\Big|_{a=\delta_p^\pi(\boldsymbol{x})}, \\ PEPL_2(\pi, \boldsymbol{x}) &= \text{E}\left[L_p(\theta, a)|\boldsymbol{x}\right]\Big|_{a=\delta_2^\pi(\boldsymbol{x})}. \end{aligned} \tag{29}$$

The power-power loss function is proposed in [11], and it has all the seven properties proposed in his paper. More specifically, it penalizes gross overestimation and gross underestimation equally, is convex in its argument, and has balanced convergence rates or penalties for its argument too large and too small. Therefore, it is recommended for the positive restricted parameter space $\Theta = (0, \infty)$.

4. Bayesian estimation for $\theta \in (0, 1)$

There are some hierarchical models where the unknown parameter of interest is $\theta \in \Theta = (0, 1)$. As pointed out in the introduction, we should calculate and use the Bayesian estimator of the parameter θ under the power-log loss function or Zhang's loss function because they penalize gross overestimation and gross underestimation equally. We list two such hierarchical models as follows.

Model (d) (beta-binomial model). This hierarchical model has been investigated by [1, 12, 13, 16, 32, 33]. Suppose that $X_1, X_2, \ldots, X_n$ are from the beta-binomial model:

$$\begin{cases} X_i|\theta \overset{\text{iid}}{\sim} Bin(m, \theta), & i = 1, 2, \ldots, n, \\ \theta \sim Be(\alpha, \beta), \end{cases} \tag{30}$$

where $\alpha > 0$ and $\beta > 0$ are known constants, m is a known positive integer, $\theta \in (0, 1)$ is the unknown parameter of interest, $Be(\alpha, \beta)$ is the beta distribution, and $Bin(m, \theta)$ is the binomial distribution. Specifically, Bayesian estimation of θ under the prior $Be(\alpha, \beta)$ is studied in Example 7.2.14 (p. 324) of [16] and in Tables 3.3.1 (p. 121) and 4.2.1 (p. 176) of [1]. However, they only calculate the Bayesian estimator with respect to $Be(\alpha, \beta)$ prior under the squared error loss, $\delta_2^\pi(\boldsymbol{x}) = \mathrm{E}(\theta|\boldsymbol{x})$. Moreover, they only consider one observation. The beta-binomial model has been investigated recently. For instance, [32] uses the beta-binomial to draw the random removals in progressive censoring; [12, 13] use the beta-binomial to model some magazine exposure data for the monthly magazine *Signature*; [33] develops estimation procedure for the parameters of a zero-inflated overdispersed binomial model in the presence of missing responses.

Model (e) (beta-negative binomial model). This hierarchical model has been investigated by [1, 34]. Suppose that $X_1, X_2, \ldots, X_n$ are from the beta-negative binomial model:

$$\begin{cases} X_i|\theta \overset{\text{iid}}{\sim} NB(m, \theta), & i = 1, 2, \ldots, n, \\ \theta \sim Be(\alpha, \beta), \end{cases} \tag{31}$$

where $\alpha > 0$ and $\beta > 0$ are known constants, m is a known positive integer, $\theta \in (0, 1)$ is the unknown parameter of interest, $Be(\alpha, \beta)$ is the beta distribution, and $NB(m, \theta)$ is the negative binomial distribution. Specifically, Bayesian estimation of θ under the prior $Be(\alpha, \beta)$ is studied in Tables 3.3.1 (p. 121) and 4.2.1 (p. 176) of [1]. However, he only calculates the Bayesian estimator with respect to $Be(\alpha, \beta)$ prior under the squared error loss function, $\delta_2^\pi(\boldsymbol{x}) = \mathrm{E}(\theta|\boldsymbol{x})$. Moreover, he only considers one observation.

4.1 Power-log loss function

A good loss function $L(\theta, a) = L(a|\theta) = L(x)|_{x=a/\theta}$ for $\Theta = (0, 1)$ should have the six properties summarized in **Table 1** (see **Table 1** in [12]).

In **Table 1**, property (a) means that any action a of the parameter θ should incur a nonnegative loss. Property (b) means that when $x = a/\theta = 1$, or $a = \theta$, that is, a correctly estimates θ, the loss is 0. Property (c) means that when $x = a/\theta \to (1/\theta)^-$, that is, a is moving away from θ and tends to 1^-, it will incur an infinite loss. Property (d) means that when $x = a/\theta \to 0^+$, that is, a is moving away from θ and tends to 0^+, it will also incur an infinite loss. Properties (c) and (d) mean that the loss function will penalize gross overestimation and gross underestimation equally. Property (e) is useful in the proofs of some propositions of the minimaxity and the admissibility of the Bayesian estimator (see [1]). Property (f) means that 1 and θ are the local extrema of $L(x)$ and $L(a|\theta)$, respectively. Property (f) also implies that $L(\theta + \Delta a|\theta) = o(\Delta a)$, that is, the loss incurred by an action $a = \theta + \Delta a$ near θ ($\Delta a \approx 0$), is very small compared to Δa.

Let

$$g_{pl}(x) = \frac{\left(\frac{1}{\theta} - 1\right)^2}{\frac{1}{\theta} - x} - \log x \text{ and } g_{pl}(1) = \frac{1}{\theta} - 1. \tag{32}$$

Define

$$L_{pl}(x) = g_{pl}(x) - g_{pl}(1) = \frac{\left(\frac{1}{\theta} - 1\right)^2}{\frac{1}{\theta} - x} - \log x - \left(\frac{1}{\theta} - 1\right). \tag{33}$$

Thus

$$\begin{aligned} L_{pl}(\theta, a) = L_{pl}(a|\theta) = L_{pl}(x)\big|_{x=a/\theta} &= \frac{\left(\frac{1}{\theta} - 1\right)^2}{\frac{1}{\theta} - \frac{a}{\theta}} - \log\frac{a}{\theta} - \left(\frac{1}{\theta} - 1\right) \\ &= \frac{\theta\left(\frac{1}{\theta} - 1\right)^2}{1 - a} - \log a + \log\theta - \left(\frac{1}{\theta} - 1\right). \end{aligned} \tag{34}$$

It is easy to check (see the supplement of [12]) that $L_{pl}(\theta, a) = L_{pl}(a|\theta) = L_{pl}(x)\big|_{x=a/\theta}$, which is called the power-log loss function, satisfies all the six properties listed in **Table 1**. Consequently, the power-log loss function is a good loss function for $\Theta = (0, 1)$, and thus it is recommended for $\Theta = (0, 1)$.

We remark that the power-log loss function on $\Theta = (0, 1)$ is an analog of the power-log loss function on $\Theta = (0, \infty)$, which is the popular Stein's loss function.

Properties	$L(x)$	$L(a\|\theta)$
(a)	$L(x) \geq 0$ for all $0 < x < \frac{1}{\theta}$	$L(a\|\theta) \geq 0$ for all $0 < a < 1$
(b)	$L(1) = 0$	$L(\theta\|\theta) = L(a\|\theta)\big\|_{a=\theta} = 0$
(c)	$L\left(\left(\frac{1}{\theta}\right)^-\right) = \lim_{x \to \left(\frac{1}{\theta}\right)^-} L(x) = \infty$	$L(1^-\|\theta) = \lim_{a \to 1^-} L(a\|\theta) = \infty$
(d)	$L(0^+) = \lim_{x \to 0^+} L(x) = \infty$	$L(0^+\|\theta) = \lim_{a \to 0^+} L(a\|\theta) = \infty$
(e)	Convex in x for all $0 < x < \frac{1}{\theta}$	Convex in a for all $0 < a < 1$
(f)	$L'(1) = \frac{dL(x)}{dx}\Big\|_{x=1} = 0$	$\left[\frac{\partial}{\partial a} L(a\|\theta)\right]\Big\|_{a=\theta} = 0$

Table 1.
(Table 1 in [12]) The six properties of a good loss function for $\Theta = (0, 1)$. $0 < \theta < 1$ is fixed.

The Bayesian estimator under the power-log loss function, $\delta_{pl}^{\pi}(\boldsymbol{x})$, minimizes the posterior expected power-log loss (PEPLL) (see [12]), $\mathrm{E}\left[L_{pl}(\theta,a)|\boldsymbol{x}\right]$, that is,

$$\delta_{pl}^{\pi}(\boldsymbol{x}) = \arg\min_{a\in\mathcal{A}} \mathrm{E}\left[L_{pl}(\theta,a)|\boldsymbol{x}\right], \tag{35}$$

where $\mathcal{A}\{a(\boldsymbol{x}) : a(\boldsymbol{x}) \in (0,1)\}$ is the action space, $a = a(\boldsymbol{x}) \in (0,1)$ is an action (estimator), $L_{pl}(\theta,a)$ given by (34) is the power-log loss function, and $\theta \in (0,1)$ is the unknown parameter of interest. The PEPLL is easy to obtain (see [12]):

$$PEPLL(\pi,a|\boldsymbol{x}) = \mathrm{E}\left[L_{pl}(\theta,a)|\boldsymbol{x}\right] = \frac{E_1(\boldsymbol{x})}{1-a} - \log a + E_2(\boldsymbol{x}) - E_3(\boldsymbol{x}) + 1, \tag{36}$$

where

$$\begin{aligned} E_1(\boldsymbol{x}) &= \mathrm{E}\left[\theta^{-1}(1-\theta)^2|\boldsymbol{x}\right] > 0, \\ E_2(\boldsymbol{x}) &= \mathrm{E}[\log\theta|\boldsymbol{x}] < 0, \\ E_3(\boldsymbol{x}) &= \mathrm{E}\left[\theta^{-1}|\boldsymbol{x}\right] > 0. \end{aligned} \tag{37}$$

It is found in [12] that

$$\delta_{pl}^{\pi}(\boldsymbol{x}) = \frac{2 + E_1(\boldsymbol{x}) - \sqrt{E_1(\boldsymbol{x})(E_1(\boldsymbol{x})+4)}}{2} \tag{38}$$

by taking partial derivative of the PEPLL with respect to a and setting it to 0. The PEPLLs evaluated at the Bayesian estimators are (see [12])

$$\begin{aligned} PEPLL_{pl}(\pi,\boldsymbol{x}) &= \mathrm{E}\left[L_{pl}(\theta,a)|\boldsymbol{x}\right]\big|_{a=\delta_{pl}^{\pi}(\boldsymbol{x})}, \\ PEPLL_2(\pi,\boldsymbol{x}) &= \mathrm{E}\left[L_{pl}(\theta,a)|\boldsymbol{x}\right]\big|_{a=\delta_2^{\pi}(\boldsymbol{x})}. \end{aligned} \tag{39}$$

Finally, the numerical simulations and a real data example of some monthly magazine exposure data (see [35]) exemplify the theoretical studies of two size relationships about the Bayesian estimators and the PEPLLs in [12].

4.2 Zhang's loss function

Zhang et al. [12] proposed six properties for a good loss function $L(\theta,a) = L(a|\theta) = L(x)|_{x=a/\theta}$ on $\Theta = (0,1)$. Apart from the six properties, [13] proposes the seventh property (balanced convergence rates or penalties for the argument too large and too small) for a good loss function on $\Theta = (0,1)$. Moreover, the seven properties for a good loss function on $\Theta = (0,1)$ are summarized in **Table 1** of [13]. The explanations of the first six properties in **Table 1** of [13] can be found in the previous subsection (see also [12]). In **Table 1** of [13], property (g) (the seventh property) means that $L\left(k_1(\theta)\frac{1}{n}\right)$ and $L\left(\frac{1}{\theta}\left(1-\frac{1}{n}\right)\right)$ tend to ∞ at the same rate and $L\left(k_2(\theta)\frac{1}{n}|\theta\right)$ and $L\left(1-\frac{1}{n}|\theta\right)$ tend to ∞ at the same rate. In other words,

$$\lim_{n\to\infty}\frac{L\left(k_1(\theta)\frac{1}{n}\right)}{L\left(\frac{1}{\theta}\left(1-\frac{1}{n}\right)\right)} = 1 \text{and} \lim_{n\to\infty}\frac{L\left(k_2(\theta)\frac{1}{n}|\theta\right)}{L\left(1-\frac{1}{n}|\theta\right)} = 1. \tag{40}$$

And they say that $L\left(k_1(\theta)\frac{1}{n}\right)$ and $L\left(\frac{1}{\theta}\left(1-\frac{1}{n}\right)\right)$ are asymptotically equivalent. Similarly, $L\left(k_2(\theta)\frac{1}{n}|\theta\right)$ and $L\left(1-\frac{1}{n}|\theta\right)$ are said to be asymptotically equivalent. They also say that $L(x)$ $(L(a|\theta))$ has balanced convergence rates or penalties for x (a) too large and too small. It is worthy to note that $k_1(\theta)\frac{1}{n}\to 0$ and $\frac{1}{\theta}\left(1-\frac{1}{n}\right)\to\frac{1}{\theta}$ at the same order $O\left(\frac{1}{n}\right)$. Analogously, $k_2(\theta)\frac{1}{n}\to 0$ and $1-\frac{1}{n}\to 1$ at the same order $O\left(\frac{1}{n}\right)$. Finally, only when properties (c) and (d) hold, property (g) may hold.

Let

$$g_z(x)=\frac{1}{\left(\frac{1}{\theta}-1\right)^2 x}+\frac{1}{\frac{1}{\theta}-x}\ \text{and}\ g_z(1)=\frac{1}{\theta\left(\frac{1}{\theta}-1\right)^2}. \tag{41}$$

Let

$$L_z(x)=g_z(x)-g_z(1)=\frac{1}{\left(\frac{1}{\theta}-1\right)^2 x}+\frac{1}{\frac{1}{\theta}-x}-\frac{1}{\theta\left(\frac{1}{\theta}-1\right)^2}. \tag{42}$$

Thus

$$\begin{aligned} L_z(\theta,a)=L_z(a|\theta)=L_z(x)|_{x=a/\theta} &= \frac{1}{\left(\frac{1}{\theta}-1\right)^2\frac{a}{\theta}}+\frac{1}{\frac{1}{\theta}-\frac{a}{\theta}}-\frac{1}{\theta\left(\frac{1}{\theta}-1\right)^2} \\ &= \frac{\theta}{\left(\frac{1}{\theta}-1\right)^2 a}+\frac{\theta}{1-a}-\frac{1}{\theta\left(\frac{1}{\theta}-1\right)^2}. \end{aligned} \tag{43}$$

It is easy to check (see the supplement of [13]) that $L_z(\theta,a)=L_z(a|\theta)=L_z(x)|_{x=a/\theta}$, which is called Zhang's loss function, satisfies all the seven properties listed in **Table 1** of [13]. Consequently, Zhang's loss function is a good loss function, and thus it is recommended for $\Theta=(0,1)$.

The Bayesian estimator under Zhang's loss function, $\delta_z^\pi(\boldsymbol{x})$, minimizes the posterior expected Zhang's loss (PEZL) (see [13]), $\mathrm{E}[L_z(\theta,a)|\boldsymbol{x}]$, that is,

$$\delta_z^\pi(\boldsymbol{x})=\arg\min_{a\in\mathcal{A}}\mathrm{E}[L_z(\theta,a)|\boldsymbol{x}], \tag{44}$$

where $\mathcal{A}\{a(\boldsymbol{x}):a(\boldsymbol{x})\in(0,1)\}$ is the action space, $a=a(\boldsymbol{x})\in(0,1)$ is an action (estimator), $L_z(\theta,a)$ given by (43) is Zhang's loss function, and $\theta\in(0,1)$ is the unknown parameter of interest. The PEZL is easy to obtain (see [13]):

$$PEZL(\pi,a|\boldsymbol{x})=\mathrm{E}[L_z(\theta,a)|\boldsymbol{x}]=\frac{E_1(\boldsymbol{x})}{a}+\frac{E_2(\boldsymbol{x})}{1-a}-E_3(\boldsymbol{x}), \tag{45}$$

where

$$\begin{aligned} E_1(\boldsymbol{x}) &= \mathrm{E}\left[\frac{\theta^3}{(1-\theta)^2}|\boldsymbol{x}\right], \\ E_2(\boldsymbol{x}) &= \mathrm{E}(\theta|\boldsymbol{x}), \\ E_3(\boldsymbol{x}) &= \mathrm{E}\left[\frac{\theta}{(1-\theta)^2}|\boldsymbol{x}\right]. \end{aligned} \tag{46}$$

It is found in [13] that

$$\delta_z^{\pi}(\boldsymbol{x}) = \frac{\sqrt{E_1(\boldsymbol{x})}}{\sqrt{E_1(\boldsymbol{x})} + \sqrt{E_2(\boldsymbol{x})}} \tag{47}$$

by taking partial derivative of the PEZL with respect to a and setting it to 0. The PEZLs evaluated at the Bayesian estimators are (see [13])

$$\begin{aligned} PEZL_z(\pi, \boldsymbol{x}) &= \mathrm{E}[L_z(\theta, a)|\boldsymbol{x}]|_{a=\delta_z^{\pi}(\boldsymbol{x})}, \\ PEZL_2(\pi, \boldsymbol{x}) &= \mathrm{E}[L_z(\theta, a)|\boldsymbol{x}]|_{a=\delta_2^{\pi}(\boldsymbol{x})}. \end{aligned} \tag{48}$$

Zhang et al. [13] considers an example of some magazine exposure data for the monthly magazine *Signature* (see [12, 35]) and compares the numerical results with those of [12].

For the probability parameter θ of the beta-negative binomial model (31), [34] recommends and analytically calculates the Bayesian estimator $\delta_z^{\pi}(\boldsymbol{x})$, with respect to $Be(\alpha, \beta)$ prior under Zhang's loss function which penalizes gross overestimation and gross underestimation equally. This estimator minimizes the PEZL. They also calculate the usual Bayesian estimator $\delta_2^{\pi}(\boldsymbol{x}) = \mathrm{E}(\theta|\boldsymbol{x})$ which minimizes the PESEL. Moreover, they also obtain the PEZLs evaluated at the two Bayesian estimators, $PEZL_z(\pi, \boldsymbol{x})$ and $PEZL_2(\pi, \boldsymbol{x})$. After that, they show two theorems about the estimators of the hyperparameters of the beta-negative binomial model (31) when m is known or unknown by the moment method (Theorem 1 in [34]) and the MLE method (Theorem 2 in [34]). Finally, the empirical Bayesian estimator of the probability parameter θ under Zhang's loss function is obtained with the hyperparameters estimated by the moment method or the MLE method from the two theorems.

In the numerical simulations of [34], they have illustrated three things: the two inequalities of the Bayesian posterior estimators and the PEZLs, the moment estimators and the MLEs, which are consistent estimators of the hyper-parameters, and the goodness of fit of the beta-negative binomial model to the simulated data. Numerical simulations show that the MLEs are better than the moment estimators when estimating the hyperparameters in terms of the goodness of fit of the model to the simulated data. However, the MLEs are very sensitive to the initial estimators, and the moment estimators are usually proved to be good initial estimators.

In the real data section of [34], they consider an example of some insurance claim data, which are assumed from the beta-negative binomial model (31). They consider four cases to fit the real data. In the first case, they assume that $m = 6$ is known for illustrating purpose (of course, one can assume another known m value). In the other three cases, they assume that m is unknown, and they provide three approaches to handle this scenario. The first two approaches consider a range of m values, for instance, $m = 1, 2, \ldots, 20$. The first approach is to maximize the log-likelihood function. The second approach is to maximize the p-value of the goodness of fit of the model (31) to the real data. The third approach is to determine the hyperparameters α, β, and m from Theorems 1 and 2 in [34] by the moment method and the MLE method, respectively, when m is unknown. Four tables which show the number of claims, the observed frequencies, the expected probabilities, and the expected frequencies of the insurance claims data are provided to illustrate the four cases.

5. Inequalities among Bayesian posterior estimators

For the six loss functions, we have the corresponding six Bayesian estimators $\delta^{\pi}_{w2}(\boldsymbol{x})$, $\delta^{\pi}_{pl}(\boldsymbol{x})$, $\delta^{\pi}_{s}(\boldsymbol{x})$, $\delta^{\pi}_{p}(\boldsymbol{x})$, $\delta^{\pi}_{2}(\boldsymbol{x})$, and $\delta^{\pi}_{z}(\boldsymbol{x})$. Interestingly, for the six Bayesian estimators, we discover three strings of inequalities which are summarized in Theorem 1 (see Theorem 1 in [36]). To our surprise, an order between the two Bayesian estimators $\delta^{\pi}_{w2}(\boldsymbol{x})$ and $\delta^{\pi}_{pl}(\boldsymbol{x})$ on $\Theta = (0, 1)$ does not exist. It is worthy to note that the three strings of inequalities only depend on the loss functions. Moreover, the inequalities are independent of the chosen models, and the used priors provided the Bayesian estimators exist, and thus they exist in a general setting which makes them quite interesting.

In this section, we compare the six Bayesian estimators $\delta^{\pi}_{w2}(\boldsymbol{x})$, $\delta^{\pi}_{pl}(\boldsymbol{x})$, $\delta^{\pi}_{s}(\boldsymbol{x})$, $\delta^{\pi}_{p}(\boldsymbol{x})$, $\delta^{\pi}_{2}(\boldsymbol{x})$, and $\delta^{\pi}_{z}(\boldsymbol{x})$. The domains of the loss functions, the six Bayesian estimators, the PELs, and the smallest PELs are summarized in **Table 2** (see **Table 1** in [36]). The six PELs are PEWSEL, PEPLL, PESL, PEPL, PESEL, and PEZL. In **Table 2**, each Bayesian estimator minimizes some corresponding PEL. Furthermore, the smallest PEL is the PEL evaluated at the corresponding Bayesian estimator.

It is easy to see that all the six loss functions are well defined on $\Theta = (0, 1)$, and thus all the six Bayesian estimators are well defined on $\Theta = (0, 1)$. There are only four loss functions defined on $\Theta = (0, \infty)$, since the power-log loss function and Zhang's loss function are only defined on $\Theta = (0, 1)$. Hence, only four Bayesian estimators are well defined on $\Theta = (0, \infty)$. Moreover, only the weighted squared error loss function and the squared error loss function are defined on $\Theta = (-\infty, \infty)$, and therefore only two Bayesian estimators are well defined on $\Theta = (-\infty, \infty)$. Among the six Bayesian estimators, there exist three strings of inequalities which are summarized in the following theorem.

Theorem 1 (Theorem 1 in [36]). *Assume the prior satisfies some regularity conditions so that the posterior expectations involved in the definitions of the six Bayesian estimators exist. Then for* $\Theta = (0, 1)$, *there exists a string of inequalities among the six Bayesian estimators*:

$$\max\left(\delta^{\pi}_{w2}(\boldsymbol{x}), \delta^{\pi}_{pl}(\boldsymbol{x})\right) \leq \delta^{\pi}_{s}(\boldsymbol{x}) \leq \delta^{\pi}_{p}(\boldsymbol{x}) \leq \delta^{\pi}_{2}(\boldsymbol{x}) \leq \delta^{\pi}_{z}(\boldsymbol{x}). \tag{49}$$

Moreover, for $\Theta = (0, \infty)$, *there exists a string of inequalities among the four Bayesian estimators*:

$$\delta^{\pi}_{w2}(\boldsymbol{x}) \leq \delta^{\pi}_{s}(\boldsymbol{x}) \leq \delta^{\pi}_{p}(\boldsymbol{x}) \leq \delta^{\pi}_{2}(\boldsymbol{x}). \tag{50}$$

Finally, for $\Theta = (-\infty, \infty)$, *there exists an inequality between the two Bayesian estimators*:

$$\delta^{\pi}_{w2}(\boldsymbol{x}) \leq \delta^{\pi}_{2}(\boldsymbol{x}). \tag{51}$$

The proof of Theorem 1 exploits a key, important, and unified tool, the covariance inequality (see Theorem 4.7.9 (p. 192) in [16]), and the proof can be found in the supplement of [36].

It is worthy to note that the six Bayesian estimators and the six smallest PELs are all functions of π, $\boldsymbol{x}$, and the loss function. Because there exists three strings of inequalities among the six Bayesian estimators, we would wonder whether there exists a string of inequalities among the six smallest PELs, in other words, $PEWSEL_{w2}(\pi, \boldsymbol{x})$, $PEPLL_{pl}(\pi, \boldsymbol{x})$, $PESL_{s}(\pi, \boldsymbol{x})$, $PEPL_{p}(\pi, \boldsymbol{x})$, $PESEL_{2}(\pi, \boldsymbol{x})$, and $PEZL_{z}(\pi, \boldsymbol{x})$. The answer to this question is no! The numerical simulations of the smallest PELs exemplify this fact (see [36]).

Domain	Bayesian estimators	PELs	Smallest PELs
$\Theta = (-\infty, \infty)$	$\delta^{\pi}_{w2}(\boldsymbol{x}) = \frac{E\left(\frac{1}{\theta}\mid\boldsymbol{x}\right)}{E\left(\frac{1}{\theta^2}\mid\boldsymbol{x}\right)}$	$PEWSEL(\pi, a\mid\boldsymbol{x}) = E[L_{w2}(\theta, a)\mid\boldsymbol{x}] = E\left[\frac{1}{\theta^2}(\theta - a)^2\mid\boldsymbol{x}\right]$	$PEWSEL_{w2}(\pi, \boldsymbol{x}) = PEWSEL(\pi, a\mid\boldsymbol{x})\mid_{a=\delta^{\pi}_{w2}(\boldsymbol{x})}$
$\Theta = (0, 1)$	$\delta^{\pi}_{pl}(\boldsymbol{x}) = \frac{2 + E^{pl}_1(\boldsymbol{x}) - \sqrt{E^{pl}_1(\boldsymbol{x})\left(E^{pl}_1(\boldsymbol{x}) + 4\right)}}{2}$ with $E^{pl}_1(\boldsymbol{x}) = E\left[\frac{(1-\theta)^2}{\theta}\mid\boldsymbol{x}\right] > 0$	$PEPLL(\pi, a\mid\boldsymbol{x}) = E[L_{pl}(\theta, a)\mid\boldsymbol{x}] = E\left[\frac{\theta\left(\frac{1}{\theta} - 1\right)^2}{1 - a} - \log a + \log\theta - \frac{1}{\theta} + 1\mid\boldsymbol{x}\right]$	$PEPLL_{pl}(\pi, \boldsymbol{x}) = PEPLL(\pi, a\mid\boldsymbol{x})\mid_{a=\delta^{\pi}_{pl}(\boldsymbol{x})}$
$\Theta = (0, \infty)$	$\delta^{\pi}_{s}(\boldsymbol{x}) = \frac{1}{E\left(\frac{1}{\theta}\mid\boldsymbol{x}\right)}$	$PESL(\pi, a\mid\boldsymbol{x}) = E[L_s(\theta, a)\mid\boldsymbol{x}] = E\left[\frac{a}{\theta} - \log\frac{a}{\theta} - 1\mid\boldsymbol{x}\right]$	$PESL_s(\pi, \boldsymbol{x}) = PESL(\pi, a\mid\boldsymbol{x})\mid_{a=\delta^{\pi}_{s}(\boldsymbol{x})}$
$\Theta = (0, \infty)$	$\delta^{\pi}_{p}(\boldsymbol{x}) = \sqrt{\frac{E(\theta\mid\boldsymbol{x})}{E\left(\frac{1}{\theta}\mid\boldsymbol{x}\right)}}$	$PEPL(\pi, a\mid\boldsymbol{x}) = E[L_p(\theta, a)\mid\boldsymbol{x}] = E\left[\frac{a}{\theta} + \frac{\theta}{a} - 2\mid\boldsymbol{x}\right]$	$PEPL_p(\pi, \boldsymbol{x}) = PEPL(\pi, a\mid\boldsymbol{x})\mid_{a=\delta^{\pi}_{p}(\boldsymbol{x})}$
$\Theta = (-\infty, \infty)$	$\delta^{\pi}_{2}(\boldsymbol{x}) = E(\theta\mid\boldsymbol{x})$	$PESEL(\pi, a\mid\boldsymbol{x}) = E[L_2(\theta, a)\mid\boldsymbol{x}] = E\left[(\theta - a)^2\mid\boldsymbol{x}\right]$	$PESEL_2(\pi, \boldsymbol{x}) = PESEL(\pi, a\mid\boldsymbol{x})\mid_{a=\delta^{\pi}_{2}(\boldsymbol{x})}$
$\Theta = (0, 1)$	$\delta^{\pi}_{z}(\boldsymbol{x}) = \frac{\sqrt{E^{z}_1(\boldsymbol{x})}}{\sqrt{E^{z}_1(\boldsymbol{x})} + \sqrt{E^{z}_2(\boldsymbol{x})}}$ with $E^{z}_1(\boldsymbol{x}) = E\left[\frac{\theta^3}{(1-\theta)^2}\mid\boldsymbol{x}\right]$ and $E^{z}_2(\boldsymbol{x}) = E[\theta\mid\boldsymbol{x}]$	$PEZL(\pi, a\mid\boldsymbol{x}) = E[L_z(\theta, a)\mid\boldsymbol{x}] = E\left[\frac{\theta}{\left(\frac{1}{\theta} - 1\right)^2 a} + \frac{\theta}{1 - a} - \frac{1}{\theta\left(\frac{1}{\theta} - 1\right)^2}\mid\boldsymbol{x}\right]$	$PEZL_z(\pi, \boldsymbol{x}) = PEZL(\pi, a\mid\boldsymbol{x})\mid_{a=\delta^{\pi}_{z}(\boldsymbol{x})}$

Table 2.
(Table 1 in [36]) The six Bayesian estimators, the PELs, and the smallest PELs.

6. Conclusions and discussions

In this chapter, we have investigated six loss functions: the squared error loss function, the weighted squared error loss function, Stein's loss function, the power-power loss function, the power-log loss function, and Zhang's loss function. Now we give some suggestions on the conditions for using each of the six loss functions. It is worthy to note that among the six loss functions, the first two loss functions are defined on $\Theta = (-\infty, \infty)$ and they penalize overestimation and underestimation equally on $(-\infty, \infty)$, and thus we recommend to use them when the parameter space is $(-\infty, \infty)$. Moreover, the middle two loss functions are defined on $\Theta = (0, \infty)$, and they penalize gross overestimation and gross underestimation equally on $(0, \infty)$, and thus we recommend to use them when the parameter space is $(0, \infty)$. In particular, if one prefers the loss function to have balanced convergence rates or penalties for its argument too large and too small, then we recommend to use the power-power loss function on $(0, \infty)$. Furthermore, the last two loss functions are defined on $\Theta = (0, 1)$, and they penalize gross overestimation and gross underestimation equally on $(0, 1)$, and thus we recommend to use them when the parameter space is $(0, 1)$. In particular, if one prefers the loss function to have balanced convergence rates or penalties for its argument too large and too small, then we recommend to use Zhang's loss function on $(0, 1)$.

For each one of the six loss functions, we can find a corresponding Bayesian estimator, which minimizes the corresponding posterior expected loss. Among the six Bayesian estimators, there exist three strings of inequalities summarized in Theorem 1 (see also Theorem 1 in [36]). However, a string of inequalities among the six smallest PELs does not exist.

We summarize three hierarchical models where the unknown parameter of interest is $\theta \in \Theta = (0, \infty)$, that is, the hierarchical normal and inverse gamma model (9), the hierarchical Poisson and gamma model (10), and the hierarchical normal and normal-inverse-gamma model (11). In addition, we summarize two hierarchical models where the unknown parameter of interest is $\theta \in \Theta = (0, 1)$, that is, the beta-binomial model (30) and the beta-negative binomial model (31).

Now we give some suggestions on the selection of the hyperparameters. One way to select the hyperparameters is through the empirical Bayesian analysis, which relies on a conjugate prior modeling, where the hyperparameters are estimated from the observations and the "estimated prior" is then used as a regular prior in the later inference. The marginal distribution can then be used to recover the prior distribution from the observations. For empirical Bayesian analysis, two common methods are used to obtain the estimators of the hyperparameters, that is, the moment method and the MLE method. Numerical simulations show that the MLEs are better than the moment estimators when estimating the hyperparameters in terms of the goodness of fit of the model to the simulated data. However, the MLEs are very sensitive to the initial estimators, and the moment estimators are usually proved to be good initial estimators.

Acknowledgements

The research was supported by the Fundamental Research Funds for the Central Universities (2019CDXYST0016; 2018CDXYST0024), China Scholarship Council (201606055028), National Natural Science Foundation of China (11671060), and MOE project of Humanities and Social Sciences on the west and the border area (14XJC910001).

Conflict of interest

The author declared no potential conflicts of interest with respect to the research, authorship, and/or publication of this article.

Author details

Ying-Ying Zhang
Department of Statistics and Actuarial Science, College of Mathematics and Statistics, Chongqing University, Chongqing, China

*Address all correspondence to: robertzhangyying@qq.com; robertzhang@cqu.edu.cn

References

[1] Robert CP. The Bayesian Choice: From Decision-Theoretic Motivations to Computational Implementation. 2nd paperback ed. New York: Springer; 2007

[2] James W, Stein C. Estimation with quadratic loss. In: Proceedings of the Fourth Berkeley Symposium on Mathematical Statistics and Probability. Vol. 1. 1961. pp. 361-380

[3] Brown LD. Inadmissibility of the usual estimators of scale parameters in problems with unknown location and scale parameters. The Annals of Mathematical Statistics. 1968;**39**:29-48

[4] Brown LD. Comment on the paper by maatta and casella. Statistical Science. 1990;**5**:103-106

[5] Parsian A, Nematollahi N. Estimation of scale parameter under entropy loss function. Journal of Statistical Planning and Inference. 1996;**52**:77-91

[6] Petropoulos C, Kourouklis S. Estimation of a scale parameter in mixture models with unknown location. Journal of Statistical Planning and Inference. 2005;**128**:191-218

[7] Oono Y, Shinozaki N. On a class of improved estimators of variance and estimation under order restriction. Journal of Statistical Planning and Inference. 2006;**136**:2584-2605

[8] Ye RD, Wang SG. Improved estimation of the covariance matrix under stein's loss. Statistics & Probability Letters. 2009;**79**:715-721

[9] Bobotas P, Kourouklis S. On the estimation of a normal precision and a normal variance ratio. Statistical Methodology. 2010;**7**:445-463

[10] Zhang YY. The bayes rule of the variance parameter of the hierarchical normal and inverse gamma model under stein's loss. Communications in Statistics-Theory and Methods. 2017;**46**: 7125-7133

[11] Zhang YY. The bayes rule of the positive restricted parameter under the power-power loss with an application. Communications in Statistics-Theory and Methods. 2019; Under review

[12] Zhang YY, Zhou MQ, Xie YH, Song WH. The bayes rule of the parameter in (0, 1) under the power-log loss function with an application to the beta-binomial model. Journal of Statistical Computation and Simulation. 2017;**87**:2724-2737

[13] Zhang YY, Xie YH, Song WH, Zhou MQ. The bayes rule of the parameter in (0, 1) under zhang's loss function with an application to the beta-binomial model. Communications in Statistics-Theory and Methods. 2019. DOI: 10.1080/03610926.2019.1565840

[14] Stein C. Inadmissibility of the usual estimator for the variance of a normal distribution with unknown mean. Annals of the Institute of Statistical Mathematics. 1964;**16**:155-160

[15] Maatta JM, Casella G. Developments in decision-theoretic variance estimation. Statistical Science. 1990;**5**: 90-120

[16] Casella G, Berger RL. Statistical Inference. 2nd ed. USA: Duxbury; 2002

[17] Lehmann EL, Casella G. Theory of Point Estimation. 2nd ed. New York: Springer; 1998

[18] Raiffa H, Schlaifer R. Applied Statistical Decision Theory. Cambridge: Harvard University Press; 1961

[19] Deely JJ, Lindley DV. Bayes empirical bayes. Journal of the

American Statistical Association. 1981; **76**:833-841

[20] Zhang YY, Wang ZY, Duan ZM, Mi W. The empirical bayes estimators of the parameter of the poisson distribution with a conjugate gamma prior under stein's loss function. Journal of Statistical Computation and Simulation. 2019. DOI: 10.1080/00949655.2019.1652606

[21] Mao SS, Tang YC. Bayesian Statistics. 2nd ed. Beijing: China Statistics Press; 2012

[22] Chen MH. Bayesian statistics lecture. Statistics Graduate Summer School. China: School of Mathematics and Statistics; Northeast Normal University: Changchun; 2014

[23] Xie YH, Song WH, Zhou MQ, Zhang YY. The bayes posterior estimator of the variance parameter of the normal distribution with a normal-inverse gamma prior under stein's loss. Chinese Journal of Applied Probability and Statistics. 2018;**34**: 551-564

[24] Dey D, Srinivasan C. Estimation of a covariance matrix under stein's loss. The Annals of Statistics. 1985;**13**: 1581-1591

[25] Sheena Y, Takemura A. Inadmissibility of non-order-preserving orthogonally invariant estimators of the covariance matrix in the case of stein's loss. Journal of Multivariate Analysis. 1992;**41**:117-131

[26] Konno Y. Estimation of a normal covariance matrix with incomplete data under stein's loss. Journal of Multivariate Analysis. 1995;**52**:308-324

[27] Konno Y. Estimation of normal covariance matrices parametrized by irreducible symmetric cones under stein's loss. Journal of Multivariate Analysis. 2007;**98**:295-316

[28] Sun XQ, Sun DC, He ZQ. Bayesian inference on multivariate normal covariance and precision matrices in a star-shaped model with missing data. Communications in Statistics-Theory and Methods. 2010;**39**:642-666

[29] Ma TF, Jia LJ, Su YS. A new estimator of covariance matrix. Journal of Statistical Planning and Inference. 2012;**142**:529-536

[30] Xu K, He DJ. Further results on estimation of covariance matrix. Statistics & Probability Letters. 2015; **101**:11-20

[31] Tsukuma H. Estimation of a high-dimensional covariance matrix with the stein loss. Journal of Multivariate Analysis. 2016;**148**:1-17

[32] Singh SK, Singh U, Sharma VK. Expected total test time and Bayesian estimation for generalized lindley distribution under progressively type-ii censored sample where removals follow the beta-binomial probability law. Applied Mathematics and Computation. 2013;**222**:402-419

[33] Luo R, Paul S. Estimation for zero-inflated beta-binomial regression model with missing response data. Statistics in Medicine. 2018;**37**:3789-3813

[34] Zhou MQ, Zhang YY, Sun Y, Sun J. The empirical bayes estimators of the probability parameter of the beta-negative binomial model under Zhang's loss function. Computational Statistics and Data Analysis. 2019; Under review

[35] Danaher PJ. A markov mixture model for magazine exposure. Journal of the American Statistical Association. 1989;**84**:922-926

[36] Zhang YY, Xie YH, Song WH, Zhou MQ. Three strings of inequalities among six bayes estimators. Communications in Statistics-Theory and Methods. 2018;**47**:1953-1961